Enhanced Access to Publicly Funded Data for Science, Technology and Innovation

This document, as well as any data and map included herein, are without prejudice to the status of or sovereignty over any territory, to the delimitation of international frontiers and boundaries and to the name of any territory, city or area.

Note by Turkey
The information in this document with reference to "Cyprus" relates to the southern part of the Island. There is no single authority representing both Turkish and Greek Cypriot people on the Island. Turkey recognises the Turkish Republic of Northern Cyprus (TRNC). Until a lasting and equitable solution is found within the context of the United Nations, Turkey shall preserve its position concerning the "Cyprus issue".

Note by all the European Union Member States of the OECD and the European Union
The Republic of Cyprus is recognised by all members of the United Nations with the exception of Turkey. The information in this document relates to the area under the effective control of the Government of the Republic of Cyprus.

Please cite this publication as:
OECD (2020), *Enhanced Access to Publicly Funded Data for Science, Technology and Innovation*, OECD Publishing, Paris, *https://doi.org/10.1787/947717bc-en*.

ISBN 978-92-64-73331-2 (print)
ISBN 978-92-64-78395-9 (pdf)

Photo credits: Cover : © OECD.

Corrigenda to publications may be found on line at: *www.oecd.org/about/publishing/corrigenda.htm*.
© OECD 2020

The use of this work, whether digital or print, is governed by the Terms and Conditions to be found at *http://www.oecd.org/termsandconditions*.

Foreword

As society and economy become increasingly knowledge-based, data become a key resource. Data-driven innovation is transforming society through far-reaching effects on resource efficiency, productivity and competitiveness. It also helps address many global challenges, such as climate and demographic changes, and scarce resources. Data-intensive science is seen as the fourth paradigm, after empirical science, theoretical science and simulation. Data also create spillover effects and positive externalities, such as socialisation and behavioural change, cultural and scientific exchange, and greater levels of trust induced by increased transparency.

Access to data is not a binary concept – rather, it can be set to different degrees of openness, depending on the community of stakeholders involved. "As open as possible, as closed as necessary" is gradually replacing the "open-by-default" mantra associated with the early days of the open-access movement. Although opening up data can help advance the science, technology and innovation (STI) agenda, this needs to be balanced against issues of costs, privacy, security, intellectual property rights (IPRs) and preventing malevolent uses. The term "enhanced access to data" is used increasingly in relation to public-sector data and captures some of these important caveats around openness.

In 2006, the OECD paved the way for good practice in providing access to data by adopting the *Recommendation of the Council concerning Access to Research Data from Public Funding*. This Recommendation remains an important and valued reference in policy making of this increasingly important area.

This report contributes to the OECD horizontal project Going Digital and is complementary to the report *Enhancing Access to and Sharing of Data: Reconciling risks and Benefits for Data Re-use across Societies*. The latter examines the broad opportunities of enhancing access to and sharing of data across societies in the context of the growing importance of artificial intelligence and the Internet of Things, and discusses how related risks and challenges can be addressed, based on the analysis of current best practices across countries.

In a complementary approach, this report focuses on such opportunities for the purpose of STI, based on insights stemming from analytical work performed by the Committee for Scientific and Technological Policy (CSTP) of the OECD in 2017 and 2018. The analysis is largely based on: i) previous OECD work on access to data and open science; ii) a survey of policy makers on the current policy practice relating to enhanced access to data for STI performed in 2017, answered by 27 delegations; iii) a joint OECD-CSTP-Global Science Forum workshop entitled "Towards New Principles For Enhanced Access To Public Data For Science, Technology And Innovation", held on 13 March 2018; and iv) case studies contributed by OECD member countries and partner economies. The case studies cover a broad spectrum of issues – from top-level national strategies and governance arrangements for open access to data or open science, to more operational projects, such as the establishment of open repositories and portals. The full text of the case studies can be found online at https://community.oecd.org/community/cstp/enhanced-data-access/. A shorter version of this report has been published as Chapter 6 of the *OECD Science, Technology and Innovation Outlook 2018*.

The report starts with the definitions of terminology, scope and objectives in Chapter 1. Chapter 2 describes international initiatives in promoting enhanced access to data in publicly funded STI. Chapter 3 continues by presenting national policies in favour of access to data. The main policy gaps hindering access to data are discussed in Chapter 4, while Chapter 5 concludes with a view of potential future developments in this policy field.

This report was declassified by the Committee for Scientific and Technological Policy on 22 September 2019 by written procedure and prepared for publication by the OECD Secretariat.

Acknowledgements

Enhanced Access to Publicly Funded Data for Science, Technology and Innovation is prepared under the aegis of the OECD Committee for Scientific and Technological Policy (CSTP). CSTP delegates contributed significantly through their responses to the survey of policy makers on the current policy practice relating to enhanced access to data for science, technology and innovation (STI) performed in 2017, answered by 27 delegations, as well as the joint European Commission/OECD International Survey on Science, Technology and Innovation Policy (STIP) Compass. CSTP delegates further co-ordinated with relevant institutions in member countries and partner economies, and provided 18 case studies of national policies in favour of access to publicly funded data for STI.

The Honourable Michael Keenan MP, Minister for Human Services and Digital Transformation, Australian Government; and Michelle Wilmers, Curation and Dissemination Manager of the Global South Research on Open Educational Resources for Development (ROER4D) project, University of Cape Town, South Africa provided valuable contributions ("In my view") to broaden and deepen the debate. Renowned experts provided valuable contributions in a joint OECD-CSTP-Global Science Forum workshop entitled "Towards New Principles for Enhanced Access to Public Data for Science, Technology and Innovation", held on 13 March 2018. The report also draws on work by the Global Science Forum on open data and open science.

Enhanced Access to Publicly Funded Data for Science, Technology and Innovation was produced under the guidance of Dominique Guellec and Alessandra Colecchia. Alan Paic is the principal author, and Sylvain Fraccola the administrative co-ordinator.

Table of contents

Foreword 3

Acknowledgements 5

Abbreviations and acronyms 9

Executive summary 11

1 Data for science, technology and innovation: Definitions, scope and objectives 13
Significance and scope of data for STI 14
Enhanced access to data: A primer 19
Objectives and structure of the report 24
References 25
Notes 27

2 International initiatives promoting enhanced access to publicly funded data for science, technology and innovation 29
Initiatives pertaining to data from publicly funded research 30
Initiatives related to public-sector information 36
References 37
Notes 39

3 Current policies in favour of enhanced access to data in OECD member countries and partner economies 41
Overview of policy initiatives promoting enhanced access to data and open science 42
Motivation for policy initiatives promoting enhanced access to data for STI 45
Initiation of policy initiatives and process 49
International influences on national policy making 59
Monitoring and evaluation of policies 61
Achievements and challenges 63
References 67
Notes 71

4 Main policy gaps hindering access to data 73
Relevance of the 2006 OECD *Recommendation of the Council concerning Access to Research Data from Public Funding* 74
Addressing challenges in access to research data 78
Synthesis of policy gaps in promoting enhanced access to data for STI 82

References	98
Notes	101

5 The future of access to data for science, technology and innovation — 103

Policy issues and implications	104
Potential future developments in this policy field	106

Tables

Table 1.1. PSI components	18
Table 1.2. Five Safes framework	20
Table 2.1. Overview of FAIR principles	35
Table 3.1. Overview of initiatives covered by the case studies	44
Table 3.2. Commonalities among data-sharing principles	60

Figures

Figure 1.1. Variations of definitions for "research data"	16
Figure 1.2. Relevance of PSI to public research	18
Figure 1.3. Stakeholder expectations from open science: Results of a 2016 European stakeholder consultation on "Science 2.0"	20
Figure 1.4. Institutional policies and barriers to promoting research-data management and/or open access to research data in European universities	23
Figure 2.1. Respondent institutions	31
Figure 2.2. Influence of the OECD *Recommendation of the Council concerning Access to Research Data from Public Funding*	31
Figure 2.3. Relevance and current policy effort dedicated to achieving the objectives of the OECD *Recommendation of the Council concerning Access to Research Data from Public Funding*	33
Figure 3.1. Scope of policy initiatives reported in the survey	42
Figure 3.2. National policies or overall strategies to encourage or mandate dissemination of and open access to research data are defined at the national level	62
Figure 3.3. Korean researcher perceptions of open (research) data	65
Figure 4.1. Assessment of the relevance of the principles from the OECD *Recommendation of the Council concerning Access to Research Data from Public Funding*	76
Figure 4.2. Assessment of challenges related to data-driven and evidence-based research	79
Figure 4.3. Sweden: Separating metadata and semantics from sensitive data	85
Figure 4.4. Creating a value proposition for data repositories	93
Figure 4.5. Data skills	95

Follow OECD Publications on:

http://twitter.com/OECD_Pubs

http://www.facebook.com/OECDPublications

http://www.linkedin.com/groups/OECD-Publications-4645871

http://www.youtube.com/oecdilibrary

http://www.oecd.org/oecddirect/

This book has...   *StatLinks*
A service that delivers Excel® files from the printed page!

Look for the *StatLinks* at the bottom of the tables or graphs in this book. To download the matching Excel® spreadsheet, just type the link into your Internet browser, starting with the *https://doi.org* prefix, or click on the link from the e-book edition.

Abbreviations and acronyms

AI	Artificial intelligence
ATT	Open Science and Research Initiative (Finland)
AUD	Australian dollar
CODATA	Committee on Data (International Council for Science)
CONACYT	National Council of Science and Technology (Mexico)
CSTP	Committee for Scientific and Technological Policy (OECD)
EOSC	European Open Science Cloud
EC	European Commission
EC-SiB	Co-ordinating team of the Biodiversity Information System (Colombia)
EU	European Union
FAIR	Findable, Accessible, Interoperable and Reusable
G7	Group of Seven
G8	Group of Eight
G20	Group of Twenty
GO FAIR	GO (Global Open) Findable, Accessible, Interoperable and Reusable
GBIF	Global Biodiversity Information Facility
GDP	Gross domestic product
GDPR	General Data Protection Regulation (European Union)
GSF	Global Science Forum (OECD)
GSIM	Generic Statistical Information Model
HEFCE	Higher Education Funding Council for England
H2020	Horizon 2020
ICT	Information and communication technology
IoT	Internet of Things
IP	Intellectual property
IPR	Intellectual property right
ISC	International Council for Science
MSIT	Ministry of Science and ICT (Korea)
NCDO	National Chief Data Officer
NIH	National Institutes of Health (United States)

NPOS	National Plan Open Science (Netherlands)
NSF	National Science Foundation (United States)
OAI-PMH	Open Archives Initiative-Protocol for Metadata Harvesting
OAIS	Open Archival Information System
OECD	Organisation for Economic Co-operation and Development
OSTP	Office of Science and Technology Policy (United States)
PSI	Public-sector information
R&D	Research and development
RCUK	Research Councils UK
RDA	Research Data Alliance
SiB	Biodiversity Information System (Colombia)
SRC	Swedish Research Council
STI	Science, technology and innovation
SURF	Collaborative Organisation for ICT in Dutch Education and Research
TDM	Text and data mining
TOP	Transparency and Openness Promotion
UCT	University of Cape Town
UK	United Kingdom
UNESCO	United Nations Educational, Scientific and Cultural Organization
US	United States
USD	US dollar
VLIR	Flemish Interuniversity Council (Belgium)
ZB	Zettabyte (1 ZB = 1 trillion gigabytes or 10^{21} bytes)

Executive summary

This report covers publicly funded data for science, technology and innovation (STI). This includes both public-sector information (PSI)[1] used for research and innovation, and data produced by publicly funded research. It does not cover private-sector data – although they can be useful to science, because they raise policy issues of a different nature, such as the discussion of what data sets need to be shared due to public interest.

The benefits from enhanced access to publicly funded data for STI include: opportunity for new scientific discoveries; reproducibility of scientific results; facilitating cross-disciplinary co-operation; economic growth through better opportunities for innovation; increased resource efficiency; improved transparency and accountability regarding disbursement of public funds; better return on public investment; securing public support for research funding; and increasing public trust in research in general. Enhanced access also furthers other government/public missions, e.g. health, energy security and transportation. However, enhanced access also bears associated risks related to privacy, intellectual property, national security and the public interest, including the protection of rare and endangered species. These risks need to be adequately communicated and responsibly managed.

International initiatives are shaping national policy agendas promoting enhanced access to and data for STI. These initiatives include the OECD *Recommendation of the Council concerning Access to Research Data from Public Funding*; the OECD *Recommendation of the Council for Enhanced Access and More Effective Use of Public Sector Information*; the European Commission's *Recommendation on access to and preservation of scientific information*; and the *Directive on the Re-use of Public Sector Information*; the FAIR (Findable, Accessible, Interoperable and Reusable) principles; open-science cloud initiatives across the globe; the outputs of the Research Data Alliance; and the GO (Global Open) FAIR initiative.

National initiatives promoting enhanced access to data for STI are widespread and fall under the following categories: i) support for research infrastructure, including repositories and portals; ii) national policies and strategies promoting open access to data (often linked to broader open-science strategies or open-government initiatives); iii) creation of governance bodies to promote open access to data; and (iv) network and collaborative initiatives aiming to facilitate open access to data. These initiatives are sometimes driven from the top down, typically by a ministry or funding agency. They can also be driven from the bottom up, by individual institutions or consortia.

Although governmental policies and strategies often cite monitoring and evaluation of policy initiatives as a priority, few monitoring schemes have been implemented. The European Commission is showing a positive example by monitoring compliance with its Recommendation on access to and preservation of scientific information, as well as with its Open Research Data Pilot and Data Management Plan.

Overall, national policies promoting enhanced access to data are in the early phases of design and implementation. Scientific communities are not equally aware about the importance of access. Even so, the policy process has considerably enhanced stakeholder awareness of both the issue's importance and

---

[1] See Chapter 1 for definitions of research data and PSI.

potential benefits, and the measures undertaken to minimise and mitigate potential risks. A consensus-building process has gradually helped overcome cultural barriers to sharing. Adhering to international standards, and contributing to shaping those standards, is another achievement of those policies. In some cases, particularly where established physical repositories or portals are concerned, the impact can be quantified in terms of the volume of data stored and the extent of stakeholders' use of the data.

Nevertheless, challenges remain. Analytical work covered in this report has identified the following issues as requiring policy attention:

1. **Balancing the benefits and risks of data sharing.** "As open as possible, as closed as necessary" is complementary to the "open-by-default" principle advocated by the open-access movement. Opening data can provide benefits in advancing the STI agenda, but these need to be balanced against the issue of costs, privacy, security, intellectual property rights (IPRs) and preventing malevolent uses. A staged approach needs to be used to enhance access to data, including through sharing them within communities of certified users; adapting the degree of certification of users to the sensitivity of the data; and creating safe environments where certified users can access sensitive datasets in controlled environments.

2. **Technical standards and practices – keeping up with the pace of technological progress.** The FAIR principles all depend on the effective development and adoption of a common technical framework. The challenge is that technology development is now far outpacing standard-setting, creating regulatory gaps. Implementing the FAIR principles is an important initiative to close this policy gap.

3. **Defining responsibility and ownership.** IPR protection is a basic condition for incentivising innovation. However, advances in technology can provide opportunities for new methodologies, such as text and data mining (TDM). Copyright regulation which excludes temporary copies of text for the sole purpose of TDM can represent an impediment to research and innovation. In the case of public-private partnerships, policy objectives should be clearly defined to expressly allow or forbid private ownership over the data derived from publicly funded research.

4. **Incentives and rewards.** Recognition and rewards are needed to encourage researchers to share data. Current academic reward systems mostly encourage the publication of scientific results and do not sufficiently value data sharing. More remains to be done to raise researchers' awareness of open-government data and enhance the appeal of sharing access to data.

5. **Business models and funding** for enhanced data sharing. Costs are most often borne by data providers, while benefits accrue to users. Although shared access to data does not necessarily mean free data, in many research fields, data are expected to be provided free of charge at the point of use.

6. **Building human capital** and institutional capabilities to manage, create, curate and reuse data is a prerequisite for advancing data sharing.

7. **Exchange of sensitive data across borders.** Sensitive datasets can be shared on a more restricted basis with trusted and certified users. Significant barriers currently exist to providing such services across borders, owing to a lack of international legal frameworks ensuring similar levels of legal protection against misuse.

1 Data for science, technology and innovation: Definitions, scope and objectives

This chapter introduces the definitions, scope and objectives of the report. It starts by defining the overall significance of data for private-sector innovation, scientific research and society at large. A primer on enhanced access to data starts with a definition, a description of opportunities arising from enhanced access to publicly funded data for science, technology and innovation, and a rationale in favour of open access to data. It concludes with the objectives and structure of the report.

Significance and scope of data for STI

As near-real-time analysis accelerates knowledge and value creation across society, data are seen as a key resource of the knowledge economy. Data-intensive scientific discovery transforming the field of science, technology and innovation (STI), and data-driven innovation is transforming society through its far-reaching effects on resource efficiency, productivity and competitiveness. It also helps address many global challenges, such as climate and demographic changes, and scarce resources. In this context, the issue of text and data mining is a hot topic. This technique allows extracting valuable knowledge and information from large digital datasets, but is restricted, subject to personal privacy and intellectual property rights (IPRs) in many countries (OECD, 2015a).

Overall significance of data to private-sector innovation, scientific research and society

Private-sector innovation

As companies obtain detailed, timely and comprehensive information about their customers, processes and employees, data themselves become a key driver of innovation (OECD, 2015a). The availability of data triggers the emergence of new products and services, such as location-based services on smartphones or emerging home automation applications based on Internet of Things (IoT) devices. Data about consumers have revolutionised marketing techniques by allowing personalised marketing, experienced daily on social networks. Data flows facilitate the establishment and operation of global value chains, driving organisational innovation. In a 2014 survey, corporate chief executive officers stated that big data would increase their operational efficiency (51%); inform strategic direction (36%); improve customer service (27%); help identify and develop new products and services (24%); and enhance customer experience (20%) (Philip Chen and Zhang, 2014).

Private actors are increasingly aware of the potential of data within the broader category of knowledge-based capital:[1] in 13 OECD member countries, companies invest more in knowledge-based capital than in physical capital (OECD, 2017a).

Scientific research

In the research sector, scientific disciplines have become increasingly data-driven thanks to the development of data-acquisition, -storage and -analysis capabilities. Scientific data are very diverse: they include observational data, which record natural phenomena (in fields such as astronomy, geoscience and demography); experimental data, which record the outcomes of man-made experiments, such as laboratory experiments in physics, chemistry and biology, or clinical trials; computational data, which are generated through large-scale simulations; and reference data, which are highly curated datasets, such as the human genome. Simulation is used to generate data based on theoretical predictions; the results are compared to actual experiments, to verify the validity of theoretical concepts and adjust them accordingly. The development of artificial intelligence (AI) will increasingly enable algorithms to detect patterns by themselves, but requires well-tended data to be trained (OECD, 2015a). In the science and technology sector, open access to data is commonly linked to open science (OECD, 2015b).

Data-intensive science is seen as the fourth paradigm (Box 1.1). Traditional science uses human intelligence to create theoretical models, which are then compared to experimental observations. Computational science uses data as the model and seeks patterns humans may not be able to detect (OECD, 2018a). Big data have already boosted the pace of discovery in disciplines such as astronomy, high-energy physics and genomics (Gordon Bell, 2009). They are spreading to environmental and health sciences (Hey, Tansley and Tolle, 2009) and are accelerating the discovery of new materials; they were even instrumental in a biochemistry discovery by Karplus, Levitt and Warshel that led to the 2013 Nobel Prize (Towns et al., 2014). Big data have the potential to make social sciences more predictive and deterministic, for example by transforming sociology into a "hard" science. The ability to exploit ubiquitous data on human behaviour – made available through tracking personal-smartphone use – could result in considerable hard data. These could enable sociologists to develop deterministic laws of human behaviour analogous to the laws of physics – thereby establishing "social physics", which could guide policy making by predicting human reactions to specific reforms (Pentland, 2014).

> **Box 1.1. Science paradigms**
>
> **Data-intensive scientific discovery as the fourth paradigm of science**
>
> 1. First paradigm (since Antiquity): empirical science, describing natural phenomena.
> 2. Second paradigm (since the Renaissance): theoretical science, constructing models and generalisations in order to establish predictions.
> 3. Third paradigm (20th century): simulations of complex phenomena.
> 4. Fourth paradigm (today): data exploration/eScience
>   - data captured by instruments or generated by simulator
>   - processed by software
>   - information/knowledge stored in computer
>   - scientist analyses database/files using data management and statistics.
>
> Source: Hey, Tansley and Tolle (2009), "The fourth paradigm: Data-intensive scientific discovery", https://www.microsoft.com/en-us/research/wp-content/uploads/2009/10/Fourth_Paradigm.pdf.

Society

Data also create spillover effects and positive externalities, such as socialisation and behavioural change, cultural and scientific exchange, and greater levels of trust induced by transparency (OECD, 2015a).

The significance of data for society will undoubtedly further increase over the next decade. The volume of data produced globally amounted to 16 zettabytes (ZB) in 2016 and is projected by International Data Corporation to grow to 163 ZB by 2025. This exponential growth will be driven by the following trends:

- The evolution of data from business background to life-critical: 20% of the data produced in 2025 will be critical to life.
- Embedded systems and the IoT will multiply both volumes and flows of data: by 2025, a connected person will interact with connected devices 400 times per day.
- Mobile and real-time data will grow: by 2025, 25% of data will be produced in real time.
- Cognitive/AI systems will change the landscape, enabling real-time analysis and decision-making.
- The need for digital security will increase (Reinsel, Gantz and Rydning, 2017).

The importance of AI is also expected to grow significantly. In pharmaceuticals, AI is set to become the primary drug-discovery tool by 2027. In materials, AI systems can use historical data to radically shorten the time needed to discover new industrial materials. In science, AI could enable novel types of discovery, based on strengths and weaknesses that complement the capabilities of human scientists (OECD, 2018a). Access to well-managed data is a key enabler of this development, as training algorithms requires large amounts of data. In image recognition, for example, AI by Microsoft and Google has achieved human-level performance after training on 1.2 million labelled images. Even though progress is expected to create less data-hungry algorithms, the progress of AI will continue to rely on large quantities of data (Simonite, 2016).

Scope of data covered: Public data for STI

The data used for science, technology and innovation (STI) fall under three broad categories:

1. public-sector information (PSI) as a broad category of information produced, curated and managed by or for government entities (Box 1.2)
2. data from publicly funded research (including data from citizen science)
3. privately owned or commercial data.

It is important to note that the distinction between PSI and publicly funded research data is often unclear. PSI is broadly defined as "information, including information products and services, generated, created, collected, processed, preserved, maintained, disseminated, or funded by or for the Government or public institution" (Box 1.2), and may include information stemming from higher education institutions and public research organisations (Table 1.1).

The recent trend is to encompass data from publicly funded research within PSI. For example, even though the original PSI Directive of the European Commission did not address research data, it has recently been extended to cover data "resulting from" publicly funded research (European Commission, 2019). The argument for this approach is that publicly funded research produces high-value datasets using public monies and should therefore be treated according to the same principles as government data.

On its data portal,[2] the US Government aggregates access to all (not only science-related) government open-data resources in one location; it provides tools and resources to conduct research, develop web and mobile applications, design data visualisations and track metrics about data usage.

However, the nature of financing can raise some issues, since most, but not all, scientific research is publicly funded. In the case of public-private partnerships, clear rules should be defined to preserve the interests of private partners. The case study contributed by Korea, for example, reports that some research data is released as government data, while some is disclosed as public-research outputs (Shin, 2018).

Research data itself are not always defined consistently, and can mean either data "resulting from" research or "used for" research, which are widely different data sets, since clearly researchers use data which go beyond data produced by research itself (Figure 1.1). The definition employed in the 2006 OECD *Recommendation of the Council concerning Access to Research Data from Public Funding* (OECD, 2006) distinguishes the broader category of "Research data" which is "used as primary sources for scientific research" and "Research data from public funding" as research data "obtained from research conducted by government agencies or departments, or conducted using public funds provided by any level of government" (Box 1.2).

Figure 1.1. Variations of definitions for "research data"

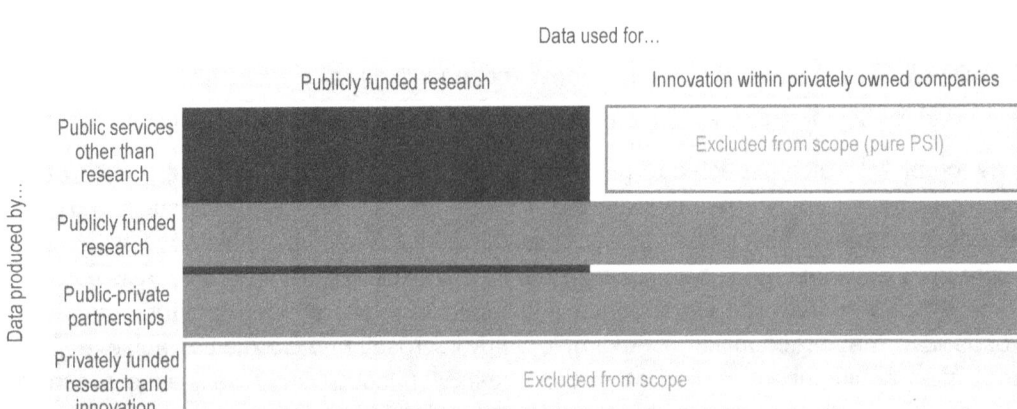

Notes: PSI = public-sector information. The light blue box represents "data for publicly funded research"; the orange box represents "data from publicly funded research". The light orange box includes data from public-private partnerships.

> **Box 1.2. Definition**
>
> **Research data**
>
> In the context of the OECD *Recommendation of the Council Concerning Access to Research Data from Public Funding* (OECD, 2006), "research data" are defined as factual records (numerical scores, textual records, images and sounds) used as primary sources for scientific research and that are commonly accepted in the scientific community as necessary to validate research findings. A research data set constitutes a systematic, partial representation of the subject being investigated.
>
> This term does not cover the following: laboratory notebooks, preliminary analyses, drafts of scientific papers, plans for future research, peer reviews, or personal communications with colleagues or physical objects (e.g. laboratory samples, bacteria strains and test animals, such as mice). Access to all of these products or outcomes of research is governed by different considerations than those dealt with here.
>
> This Recommendation principally concerns research data in a digital, computer-readable format. It is indeed in this format that the greatest potential lies for improvements in the efficient distribution of data and their application to research because the marginal costs of transmitting data through the Internet are close to zero. The Principles within the Recommendation could also apply to analogue research data in situations where the marginal costs of giving access to such data can be kept reasonably low.
>
> **Research data from public funding**
>
> Research data from public funding is defined as the research data obtained from research conducted by government agencies or departments, or conducted using public funds provided by any level of government.
>
> Given that the nature of "public funding" of research varies significantly from one country to another, the Recommendation recognises that such differences call for a flexible approach to improving access to research data (OECD, 2006).
>
> **PSI**
>
> PSI is broadly defined as "information, including information products and services, generated, created, collected, processed, preserved, maintained, disseminated, or funded by or for the Government or public institution" (OECD, 2008). Table 1.1 describes the different components of PSI.

In addition, there is a lack of consensus about the relevant depth to which research data should be made open. A first category concerns data directly underpinning the scientific results published in journals – access to this information is critical for the reproducibility of the scientific findings. Beyond that first level, there are further layers of data and intermediate results, all the way to the raw data which were initially collected, and the hypotheses which have guided the data collection.

Such a decision is also linked to the availability of algorithms and workflows needed to analyse the data. Raw data are difficult to reuse if the analysis software is not disclosed at the same time. This in turn raises issues of the capacity of other researchers to master the overall workflow and reproduce the final result. A sophisticated example was provided by CERN, who offered the possibility to re-discover the Higgs Boson using a simplified version of the original datasets, and providing the adequate software to analyse it (Jomhari, Heiser and Bin Annuar, 2017). The European Commission Open Research Pilot requires publication of the data needed to validate the results presented in scientific publications; other data can also be included by the beneficiaries on a voluntary basis.

The use of PSI for innovation is well established. A recent meta-analysis shows that innovation is the most prominent destination of open-government data utilisation. This applies to both business-driven innovation aiming to create economic value and innovation in public services (Safarov, Meijer and Grimmelikhuijsen,

2017); 73% of respondents to a European public consultation agreed that PSI increasingly provided a basis for innovative services and products (European Commission, 2017).

Respondents to a 2017 survey by the OECD Committee for Scientific and Technological Policy (CSTP) were asked to evaluate the relevance of broader sources of PSI to scientists. Figure 1.2 summarises the survey results; it shows that 74% of respondents assessed the relevance of PSI to public research as "high" or "very high".

Figure 1.2. Relevance of PSI to public research

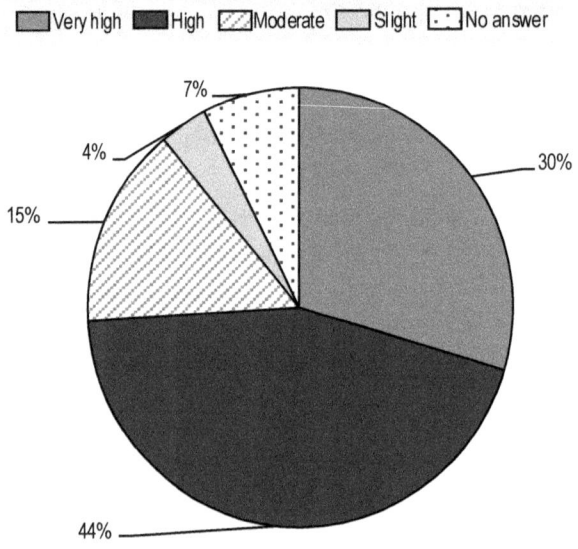

Source: Survey results from OECD and partner delegations.

StatLink: https://doi.org/10.1787/888934112519

Table 1.1. PSI components

Type of data	Examples
Geographic information	Maps, spatial and topographic data, cadastre, boundaries
Meteorological and environmental information	Meteorological, hydrographic, atmospheric, oceanographic, environmental-quality data
Economic and business information	Financial, company, industry and trade information
Social information	Demographic, health, education, labour data, attitude surveys
Traffic and transport information	Transport networks, transport and traffic, vehicle-registration data
Tourism and leisure information	Tourism statistics, hotel and entertainment data
Agriculture, farming, forestry and fisheries	Cropping/land use, farm incomes, fish farming and livestock data
Natural-resource information	Energy and natural-resource stock and consumption, biodiversity and geological data
Legal information	Crime/conviction data, legislation, jurisprudence, patent and trademark data
Scientific-research information	Information stemming from higher education institutions and public research organisations
Educational content	Academic papers and studies, lecture materials
Political content	Government press releases, proceedings, green papers
Cultural content	Museum material, archaeological sites, library resources, public archives

Source: OECD (2006), *Recommendation of the Council concerning Access to Research Data from Public Funding*, https://legalinstruments.oecd.org/en/instruments/OECD-LEGAL-0347.

Open-government data are increasingly used as inputs to scientific research. A pioneering empirical study found that researchers use open-government data from 96 different open-government portals – principally the UK and US government portals, but also a number of open-government portals in emerging countries, such as India and Kenya (Yan and Weber, 2018).

Governments are both producers and users of data; these can be used by the governments themselves, or by researchers, businesses and citizens. Governments can also use big-data techniques to understand and predict citizens' behaviour. Data are used in policy implementation, for example to model the effects of policies on human behaviour, thereby optimising policy making to achieve predictable and favourable outcomes (Pentland, 2014).

This report clearly focuses on publicly funded data. It does not address private-sector data, even though such data evidently play a large role in science and innovation. The following section reviews PSI and data from publicly funded research.

Enhanced access to data: A primer

Definition of enhanced access to data

The *OECD Principles and Guidelines for Access to Research Data from Public Funding* define access arrangements as "the regulatory, policy and procedural framework established by research institutions, research-funding agencies and other partners involved, to determine the conditions of access to and use of research data" (OECD, 2006).

It is important to note that access to data is not a binary concept – rather, it can be staged to different degrees of openness, depending on the community of stakeholders involved (OECD, 2015a). "As open as possible, as closed as necessary" is gradually replacing the "open-by-default" mantra associated with the early days of the open-access movement. Although opening up data can help advance the STI agenda, this needs to be balanced against issues of costs, privacy, security, IPRs and preventing malevolent uses. The term "enhanced access to data" is increasingly used in relation to public-sector data and captures some of these important caveats around openness.

The more sensitive the data, the more difficult it is to open them to the general public, with the underlying risk of privacy breaches and malevolent use. Hence, different degrees of openness may include: i) open access with open licence; ii) public access with a specific licence that limits use; iii) group-based access through authentication; and iv) named access explicitly assigned by contract (OECD, 2019).

A recent survey of Australian research-data repositories shows that 86% provide open access to at least part of the data, 12% offer exclusively restricted access, and 2% propose a combination of closed and restricted access. Out of the 86% repositories that are at least partly open, 50% are fully open, 32% have restricted parts of the datasets, 6% have embargoed datasets, and 6% have closed datasets (Kindling et al., 2017).[3]

Open data can be defined simply as "data that can be accessed and reused by anyone without technical or legal restrictions" (OECD, 2015b). This does not necessarily mean the data are free of cost, although in the context of open science, it is normally assumed the user bears no charges. Different models include institutional subscription to research databases; open access in the "author pays" variant – authors or their employers pay for the cost of publishing in order to provide free access to the community; open-access archives and repositories, where organisations support institutional repositories and/or subject archives, and authors make their work freely available to anyone with Internet access; and a number of hybrid solutions, such as delayed open access and open choice (Houghton and Sheehan, 2009; OECD, 2017b).

More restricted access to data can be organised within the framework of safe environments, such as the Five Safes framework (Table 1.2). These environments rely on specific safe-software platforms, where only approved researchers can access the data within a specific environment, analyse them without extracting the actual sensitive data and then submit the results of their research for approval. The results will then be investigated to test whether they risk disclosure. If they are considered "safe", the researchers will be authorised to use them; if they are considered unsafe, the researchers will need to devise a way of further anonymising the result.

1. DATA FOR SCIENCE, TECHNOLOGY AND INNOVATION: DEFINITIONS, SCOPE AND OBJECTIVES

Table 1.2. Five Safes framework

Safe projects	Is this use of the data appropriate, lawful, ethical and sensible?
Safe people	Can the researchers be trusted to use it in an appropriate manner?
Safe data	Does the data itself contain sufficient information to allow confidentiality to be breached?
Safe settings	Does the access facility limit unauthorised use or mistakes?
Safe outputs	Are the statistical results non-disclosive?

Source: Office of National Statistics UK (n.d.), "Secure research service", webpage, https://www.ons.gov.uk/aboutus/whatwedo/statistics/requestingstatistics/approvedresearcherscheme.

Opportunities from enhanced access to publicly funded data for STI

Enhanced data access and use offer opportunities for individuals, businesses and governments alike. Individuals are able to access valuable services at a negligible cost. For example, using the satellite navigation system included in any smartphone at virtually no marginal cost provides a service that is superior to the previous – and costly – option of buying a paper map and painstakingly finding one's way. Such services are based on geospatial data collected by the public sector.

Figure 1.3. Stakeholder expectations from open science: Results of a 2016 European stakeholder consultation on "Science 2.0"[1]

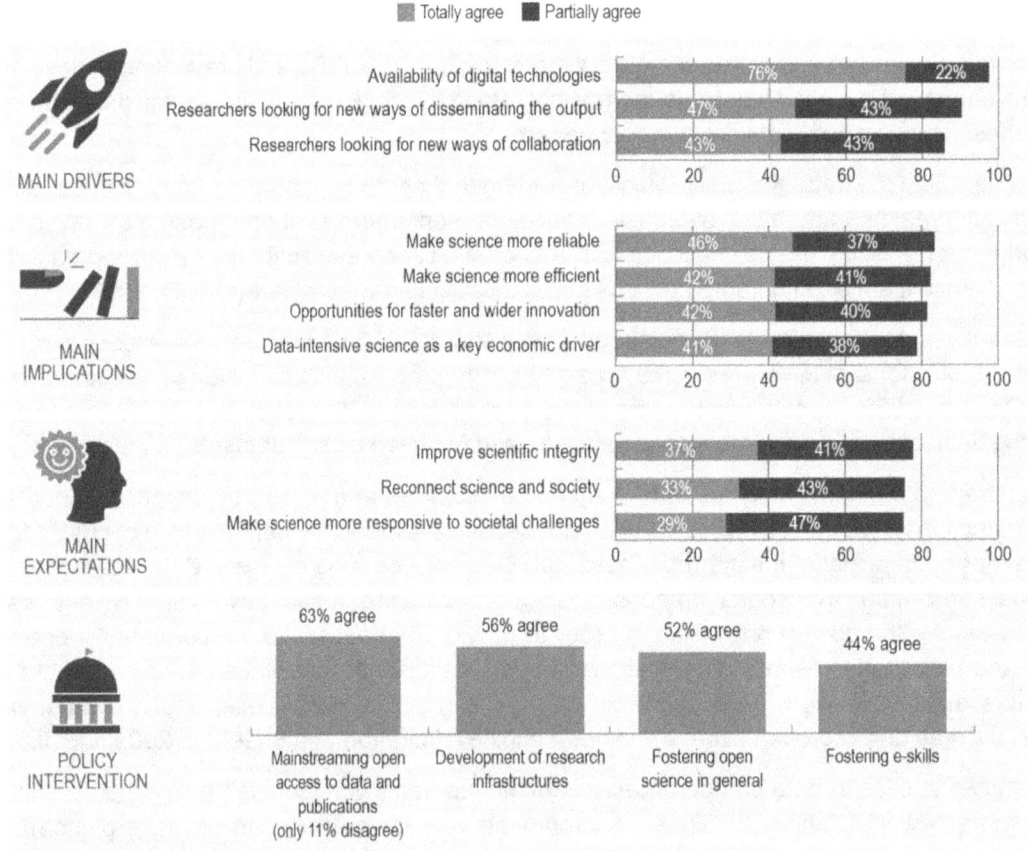

1. "Science 2.0" was used as a generic term used in 2014 to designate the next generation of science. Since then, "open science" has become the standard term.
Note: Stakeholders identified most with "open science" when referring to the future of science.
Source: European Commission (2014), "Validation of the results of the public consultation on science 2.0: Science in transition", http://ec.europa.eu/research/consultations/science-2.0/science_2_0_final_report.pdf.

Businesses use data to learn about consumer preferences, create new products and services, streamline their business processes and increase overall productivity. Moreover, data themselves are the commodity being sold. An estimate of the impact of PSI, commissioned by the European Commission, has estimated the aggregate economic impact of PSI at about EUR 140 billion in 2008, or about 1.1% of EU27 gross domestic product (GDP) (Vickery, 2011).

In the field of science and technology, open access to data is commonly linked to open science and open innovation (OECD, 2015b). An extensive European stakeholder consultation conducted in 2014 showed that open access to publications and open access to data within the context of open science were the top issues requiring policy intervention, ahead of research infrastructure and research quality. Open science is expected to make science more accountable, more efficient and more responsive to societal needs (Figure 1.3) (European Commission, 2014).

A well-known issue in science is publication bias, whereby negative results or results that are not deemed sufficiently significant are not published, because such publication is not worth the time and effort of the researchers, who will receive little or no recognition for a "non-result". However, failure to publish such data causes additional time and effort elsewhere, as such experiments may be duplicated because it is not known that avenue of research does not lead to positive discovery (Rothstein, Sutton and Borenstein, 2005). The adoption of open access to scientific data would help resolve this bias.

Drivers and barriers for open access to data

Provided that legitimate concerns about privacy, intellectual property (IP), national security and other public interests are addressed,[4] enhanced access to public data can provide great benefits to the economy, the research community and society at large. The economic benefit of enhanced access to data is quite significant, amounting to 1% or more of GDP (Box 1.3).

Box 1.3. Impact of open access to data

Estimates of the economic impact of enhanced access to data vary.

- The McKinsey Global Institute estimated the potential value creation from open access to data in seven sectors to USD 3 trillion to USD 5 trillion, or 4% to 7% of global GDP (Manyika et al., 2016).
- The OECD estimates the aggregate economic impact of PSI-related applications in OECD member countries at around USD 500 billion in 2008, equivalent to about 1.1% of cumulated GDP (OECD, 2015a).
- An Australian study estimated that data from research alone amounted to 0.15% to 0.4% of GDP in 2012, with potential upsides to 0.3% to 1% of GDP (Houghton, 2014).

At least seven main rationales exist in favour of enhanced access to publicly funded data:

1. Create opportunities for new scientific insights: Data reuse increases the efficiency of science and optimises impact and return on investments. For instance, there are more papers published using data retrieved from the archive of the Hubble Space Telescope than by the people who originally proposed and analysed observations. Providing broader access to data allows more researchers (and citizens) to analyse and link those data to other data sources in order to respond to different scientific questions. For example, the health-research community working on emerging diseases is increasingly relying on biodiversity data. Enhancing access to and sharing of data also encourages meta-analysis, which combines the results of different related studies (e.g. clinical trials of a drug) to provide greater statistical power.

2. Promote innovation and economic growth: allowing commercial companies to access and use public-research data accelerates innovation on products (e.g. new drugs) or new data services (e.g. weather forecasting). Data are the essential enabler for AI and related innovations.

3. Enhance social welfare for individuals and society at large: publicly funded research is a public good, therefore data from publicly funded research should, in principle, be available to researchers, citizens and commercial actors who wish to use and derive value from them. Transparency and accountability are sometimes an issue.

4. Increase reproducibility of scientific results: sharing access to the data underpinning scientific publications allows peers to test and reproduce scientific results. In practice, data alone are often insufficient to test reproducibility, and enhanced access to analysis software is also necessary.

5. Enhance education and training: enhanced access to data provides opportunities for richer educational content.

6. Avoid duplication: sharing datasets leading to positive or negative results can prevent duplication of research efforts (Rothstein, Sutton and Borenstein, 2005).

7. Improve governance in public research: open access to data can promote transparency, democratic accountability, citizen empowerment, better delivery of public research, innovation and use of crowd wisdom, as well as prevent duplication of data-collection efforts, optimise administrative processes and enhance access to external problem-solving capacity (Janssen, Charalabidis and Zuiderwijk, 2012). Taken together, these rationales provide for a better STI ecosystem and contribute to society as a whole. Access to data alone is not sufficient to achieve all these expectations, but lack of access is a major barrier to achieving them.

Simultaneously, enhanced access to data introduces legitimate concerns about privacy, IPRs, national security and other public interests, such as the protection of rare and endangered species. Chapter 4 addresses these risks. When, how and under what conditions public-research data should be made accessible are important policy questions that cut across the issues discussed in this report.

Progress towards achieving enhanced access to data has been uneven: the latest edition of the Open Data Barometer shows that only 7% of the data are fully open, with many fragmented, incomplete and outdated datasets, which also lack the necessary metadata (World Wide Web Foundation, 2017). Open access-to-data catalogues or portals are informally maintained; the most complete datasets are often found in other sources than the official open-data portal. Some categories of datasets are particularly important for innovation, such as map data, public transport timetables, international trade data and crime data, which entrepreneurs can use to provide specific services to end users. The degree of openness is also low in these categories (only 8% to 11% of all datasets are open) and is reported to be declining (World Wide Web Foundation, 2017).

The Global Open Data Index presents similar conclusions: i) data are hard (or even impossible) to find owing to insufficient indexation; ii) data are not readily exploitable, owing to non-standard formats, lack of machine readability (e.g. stemming from use of the HTML format) and failure to publish the raw data favoured by topical experts; and iii) open licensing is rare and jeopardised by a lack of standards, risk aversion and fear of unlawful data use, leading to ambivalent or unclear clauses that create incompatibilities between licences and hamper data use (Lämmerhirt, Rubinstein and Montiel, 2017).

The OECD Open-Useful-Reusable Government Index (OURdata Index) identifies: i) implementation gaps in late adopters of open-government data policies; ii) a need to strengthen support for reuse, both outside the public sector (through data-awareness initiatives, hackathons and co-creation events) and inside the public sector (through information sessions and regular training for civil servants); iii) an opportunity to develop platforms that allow users to actively monitor data quality and add to available data; and iv) the need to better monitor the impact of open-government data (Lafortune and Ubaldi, 2018).

1. DATA FOR SCIENCE, TECHNOLOGY AND INNOVATION: DEFINITIONS, SCOPE AND OBJECTIVES | 23

Figure 1.4. Institutional policies and barriers to promoting research-data management and/or open access to research data in European universities

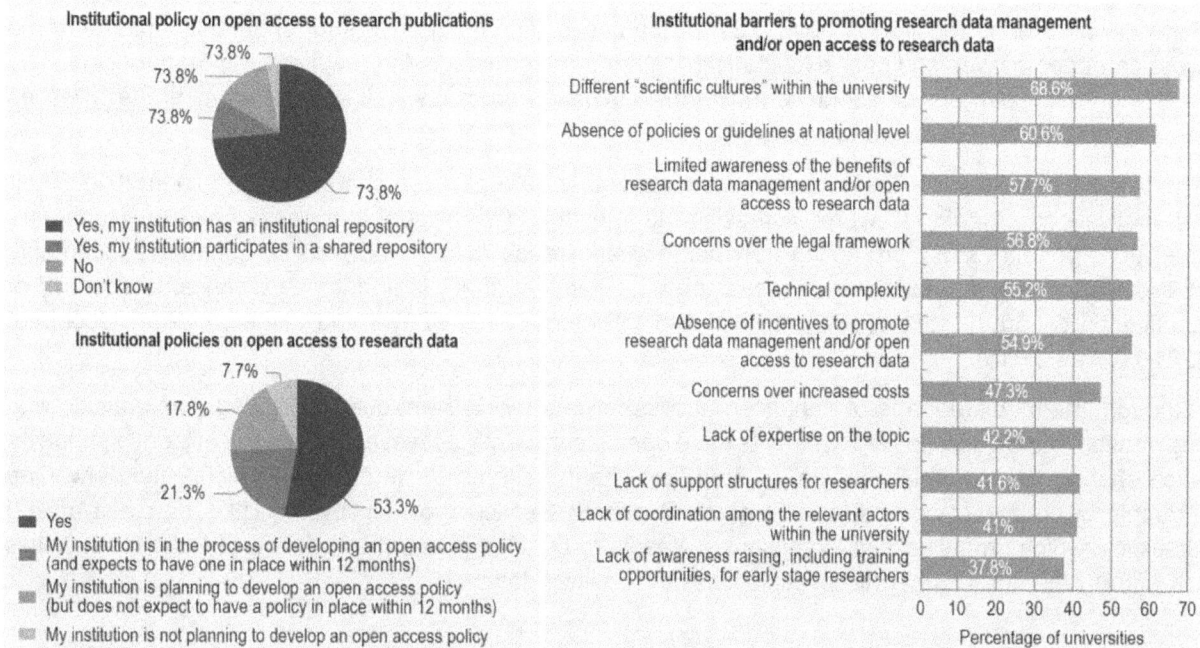

Source: Morais and Borrell-Damian (2018), "Open access – 2016-2017 EUA Survey results", www.eua.be/Libraries/publications-homepage-list/open-access-2016-2017-eua-survey-results.

In the area of research, access to data currently lags behind access to publications: although more than 92% of universities in Europe have open-access policies for publications in place or plan to have them in the near future, fewer than 28% have open access to data guidelines in place (Morais and Borrell-Damian, 2018). This is clearly not an infrastructure issue: over 83% of institutions either have their own repository or participate in a shared repository; 65% have their repository aggregated by the OpenAIRE portal/infrastructure, which aims to link the aggregated research publications to the accompanying research and project information, to enhance the reproducibility of scientific results (OpenAire, 2016). Institutional barriers to promoting research-data management include internal factors (e.g. different "scientific cultures"), limited awareness of the benefits of research data and structural elements (such as the absence of policy guidelines at the national level, the lack of incentives to promote research data and increased costs), as well as the lack of adequate infrastructure (Figure 1.4).

A 2016 OECD survey of scientific authors revealed that only 20% to 25% of corresponding authors had been asked to share data after publication. If asked, a significant number of authors (30% to 50%) said they would grant access to the data, or at least take steps to grant access, and about 30% said they would seek to clarify the request. Depending on the discipline, between 10% and 20% of authors would refuse to share data on legal grounds (Boselli and Galindo-Rueda, 2016). Authors of scientific papers are more reluctant to share their data openly than to obtain access to data from other research groups (Elsevier and CSTS, 2017).

The most recent edition of the OECD International Survey of Scientific Authors (ISSA2) shows that on average, 67% of scientific production results in new data or code (in about 24% cases it is both). Out of these, an average of about 40% get stored in repositories, and this varies from a high of over 50% for multidisciplinary research, agricultural and biological sciences and material sciences to a low of 20% to 25% in sociology, psychology, business and management. Authors seem to be more likely to share their data than code. Code was archived on a repository or delivered to a journal as supportive material in about

20% of cases, whereas around 45% of authors shared their data using these means. Re-use is yet another barrier to be overcome, since even when shared, data is not always FAIR accompanied by relevant metadata nor compliant with relevant standards, and even fewer are the cases where an object identifier is assigned. Payment of a fee is required in about 12% of the cases. The main drivers for sharing of data identified were career objectives and peer expectations, rather than formal sharing requirements from funders. The most significant barriers identified were high dissemination costs, as well as intellectual property issues (Bello and Galindo-Rueda, forthcoming).

The Wiley Open Science Researcher Insights show that 69% of researchers shared data in 2016. Their top motivations for doing so were: increasing the impact and visibility of their research (39%); furthering the public benefit (35%); promoting transparency and reuse (31%); and meeting journal requirements (29%). Conversely, the top reasons researchers hesitated to share their data were concerns about IP or confidentiality (50%), ethics (31%), misuse and misinterpretation (23%) and research being "scooped" (22%) (Wiley, 2016).

Although intermediate forms of enhanced access to data within semi-open or closed communities with registration and certification requirements have been insufficiently assessed, anecdotal evidence suggests such sharing remains relatively limited, and that sharing across borders faces particular barriers. One example of controlled access is the Secure Research Service provided by the UK Office of National Statistics, which grants access to sensitive datasets to certified researchers (Office of National Statistics UK [n.d.]). Chapter 4 further discusses access to sensitive data.

Objectives and structure of the report

This report focuses on enhanced access to publicly funded research data for STI. The objective is to take stock of current policy practices, achievements and challenges, and identify outstanding policy issues to be addressed by policymakers in the future.

Chapter 2 reviews international initiatives promoting enhanced access to data for STI. These include initiatives at the intergovernmental level – such as the OECD, the Group of Eight (G8), the European Union and UNESCO initiatives – and community-driven initiatives – such as Committee on Data of the International Council for Science, the FAIR initiative and the Research Data Alliance.

Chapter 3 reviews current policies promoting enhanced access to data in OECD member countries and partner economies. It is based on responses to the 2017 CSTP survey provided by the committee's delegates, responses to the 2017 EC/OECD STI Policy survey (EC/OECD, 2018) and the policy case studies contributed by 18 countries in 2018. The full text of the case studies is accessible on the dedicated website (OECD, 2018b).

Chapter 4 develops the main policy issues hindering enhanced access to and sharing of data: i) Balancing the benefits and risks of data sharing; ii) technical standards and practices – keeping up with the pace of technological progress; iii) defining responsibility and ownership; iv) providing incentives and rewards for data authors and stewards; v) developing business models and funding for enhanced access; vi) building human capital and institutional capabilities to manage, create, curate and reuse data; and vii) exchanging sensitive data across borders.

Finally, Chapter 5 draws conclusions from the preceding work and proposes scenarios for the future of access to data for STI.

References

Bello, M. and F. Galindo-Rueda (forthcoming), "Charting the digital transformation of science: Findings from the 2018 OECD International Survey of Scientific Authors (ISSA2)", *OECD Science, Technology and Industry Working Papers*, OECD Publishing, Paris.

Boselli, B. and F. Galindo-Rueda (2016), "Drivers and implications of scientific open access publishing: Findings from a pilot OECD International Survey of Scientific Authors", *OECD Science, Technology and Industry Policy Papers*, No. 33, OECD Publishing, Paris, http://dx.doi.org/10.1787/5jlr2z70k0bx-en.

Corrado, C., C. Hulten and D. Sichel (2004), "Measuring capital and technology: An expanded framework", https://www.federalreserve.gov/pubs/feds/2004/200465/200465pap.pdf (accessed on 26 July 2019).

EC/OECD (2018), "STIP Compass", https://stip.oecd.org/stip.html (accessed on 9 March 2020).

Elsevier and CSTS (2017), "Open data: The researcher perspective", https://www.elsevier.com/__data/assets/pdf_file/0004/281920/Open-data-report.pdf (accessed on 9 March 2020).

European Commission (2019), "Directive (EU) 2019/1024 of the European Parliament and of the Council of 20 June 2019 on open data and the re-use of public sector information", https://eur-lex.europa.eu/legal-content/EN/TXT/?qid=1561563110433&uri=CELEX:32019L1024 (accessed on 26 July 2019).

European Commission (2017), "Consultation on PSI Directive review", http://ec.europa.eu/newsroom/dae/document.cfm?doc_id=51544 (accessed on 28 February 2020).

European Commission (2014), "Validation of the results of the public consultation on science 2.0: Science in transition", http://ec.europa.eu/research/consultations/science-2.0/science_2_0_final_report.pdf (accessed on 26 February 2020).

Gordon Bell, T. (2009), "Beyond the data deluge", *Science*, Vol. 323/5919, pp. 1297-8, http://dx.doi.org/10.1126/science.1170411.

Hey, T., S. Tansley and K. Tolle (2009), "The fourth paradigm: Data-intensive scientific discovery", Microsoft Research, https://www.microsoft.com/en-us/research/wp-content/uploads/2009/10/Fourth_Paradigm.pdf (accessed on 28 November 2019).

Houghton, J. (2014), "Open research data report to the Australian National Data Service (ANDS)", https://www.ands.org.au/__data/assets/pdf_file/0019/393022/open-research-data-report.pdf (accessed on 20 December 2019).

Houghton, J. and P. Sheehan (2009), "Estimating the potential impacts of open access to research findings", *Economic Analysis and Policy*, Vol. 39, No. 1, March, https://doi.org/10.1016/S0313-5926(09)50048-3.

Janssen, M., Y. Charalabidis and A. Zuiderwijk (2012), "Benefits, adoption barriers and myths of open data and open government", *Information Systems Management*, Vol. 29/4, pp. 258-268, http://dx.doi.org/10.1080/10580530.2012.716740.

Jomhari, N., A. Heiser and A.A. Bin Annuar (2017), "Higgs-to-four-lepton analysis example using 2011-2012 data", CERN Open Data Portal, http://dx.doi.org/10.7483/OPENDATA.CMS.JKB8.RR42.

Kindling, M. et al. (2017), "The landscape of research data repositories in 2015: A re3data analysis", *D-Lib Magazine*, Vol. 23/3/4, http://dx.doi.org/10.1045/march2017-kindling.

Lafortune, G. and B. Ubaldi (2018), "OECD 2017 OURdata Index: Methodology and results", *OECD Working Papers on Public Governance*, No. 30, OECD Publishing, Paris, https://dx.doi.org/10.1787/2807d3c8-en.

Lämmerhirt, D., M. Rubinstein and O. Montiel (2017), "The state of open government data in 2017 – Creating meaningful open data through multi-stakeholder dialogue", https://blog.okfn.org/files/2017/06/FinalreportTheStateofOpenGovernmentDatain2017.pdf (accessed on 14 December 2019).

Manyika, J. et al. (2016), "Digital globalization: The new era of global flows", report, *McKinsey Global Institute*, https://www.mckinsey.com/business-functions/mckinsey-digital/our-insights/digital-globalization-the-new-era-of-global-flows (accessed 24 February 2020).

Morais, R. and L. Borrell-Damian (2018), "Open access – 2016-2017 EUA Survey results", report, European University Association, www.eua.be/Libraries/publications-homepage-list/open-access-2016-2017-eua-survey-results (accessed on 19 June 2019).

OECD (2019), *Enhanced Access to and Sharing of Data: Reconciling Risks and Benefits for Data Re-use across Societies*, OECD Publishing, Paris, https://doi.org/10.1787/276aaca8-en.

OECD (2018a), *AI: Intelligent Machines, Smart Policies*, Conference Summary, OECD Digital Economy Papers, No. 270, OECD Publishing, Paris, https://doi.org/10.1787/f1a650d9-en.

OECD (2018b), "Enhanced Access to Publicly Funded Data for Science, Technology and Innovation", webpage, OECD, Paris, https://community.oecd.org/community/cstp/enhanced-data-access (accessed on 9 January 2020).

OECD (2017a), *OECD Science, Technology and Industry Scoreboard*, OECD Publishing, Paris, https://doi.org/10.1787/9789264268821-en.

OECD (2017b), "Business models for sustainable research data repositories", *OECD Science, Technology and Industry Policy Papers*, No. 47, OECD Publishing, Paris, https://doi.org/10.1787/302b12bb-en.

OECD (2015a), *Data-Driven Innovation – Big Data for Growth and Well-Being*, OECD Publishing, Paris, https://dx.doi.org/10.1787/9789264229358-en.

OECD (2015b), "Making open science a reality", *OECD Science, Technology and Industry Policy Papers*, No. 25, OECD Publishing, Paris, http://dx.doi.org/10.1787/5jrs2f963zs1-en.

OECD (2008), *Recommendation of the Council for Enhanced Access and More Effective Use of Public Sector Information*, OECD, Paris, https://legalinstruments.oecd.org/en/instruments/OECD-LEGAL-0362.

OECD (2006), *Recommendation of the Council concerning Access to Research Data from Public Funding*, OECD, Paris, https://legalinstruments.oecd.org/en/instruments/OECD-LEGAL-0347.

Office of National Statistics UK (n.d.), "Secure research service", webpage, https://www.ons.gov.uk/aboutus/whatwedo/statistics/requestingstatistics/approvedresearcherscheme.

OpenAire (2016), "OpenAIRE's mission and vision", webpage, https://www.openaire.eu/mission-and-vision (accessed on 19 December 2019).

Pentland, A. (2014), *Social Physics: How Good Ideas Spread-the Lessons from a New Science*, Scribe Publications Pty Limited, Melbourne, London.

Philip Chen, C. and C. Zhang (2014), "Data-intensive applications, challenges, techniques and technologies: A survey on Big Data", *Information Sciences*, Vol. 275, pp. 314-347, Elsevier, http://dx.doi.org/10.1016/J.INS.2014.01.015.

Reinsel, D., J. Gantz and J. Rydning (2017), "Data age 2025: The digitization of the world – From edge to core", https://www.seagate.com/files/www-content/our-story/trends/files/Seagate-WP-DataAge2025-March-2017.pdf (accessed on 13 July 2019).

Rothstein, H., A. Sutton and M. Borenstein (2005), "Publication bias in meta-analysis", in *Publication Bias in Meta-Analysis: Prevention, Assessment and Adjustments*, Ch. 1, John Wiley & Sons, Ltd, Hoboken, NJ, http://dx.doi.org/10.1002/0470870168.ch1.

Safarov, I., A. Meijer and S. Grimmelikhuijsen (2017), "Utilization of open government data: A systematic literature review of types, conditions, effects and users", *Information Polity*, Vol. 22/1, pp. 1-24, http://dx.doi.org/10.3233/IP-160012.

Shin, E. (2018), "Korean case report on enhanced access to research data", case study for the OECD project on enhanced access to data, https://community.oecd.org/servlet/JiveServlet/downloadBody/141310-102-4-263210/korean%20case%20report.pdf.

Simonite, T. (2016), "Algorithms that learn with less data could expand AI's power", MIT Technology Review, 24 May, Boston, https://www.technologyreview.com/s/601551/algorithms-that-learn-with-less-data-could-expand-ais-power/ (accessed on 28 February 2020).

Towns, J. et al. (2014), "XESDE: Accelerating Scientific Discovery" in *Computing in Science & Engineering*, Volume 16, No. 5, IEEE, Sept-Oct., https://doi.org/10.1109/MCSE.2014.80.

Vickery, G. (2011), "Review of recent studies on PSI Re-use and related market developments", https://ec.europa.eu/newsroom/dae/document.cfm?doc_id=1093 (accessed on 24 February 2020).

Wiley (2016), "Wiley global data sharing infographic", June 2017, *Wiley Open Science Researcher Insights Survey*, https://authorservices.wiley.com/asset/photos/licensing-and-open-access-photos/Wiley%20Global%20Data%20Sharing%20Infographic%20June%202017.pdf (accessed on 17 December 2019).

World Wide Web Foundation (2017), "Open data barometer global report", fourth edition, https://opendatabarometer.org/doc/4thEdition/ODB-4thEdition-GlobalReport.pdf (accessed on 24 February 2020).

Yan, A. and N. Weber (2018), "Mining open government data used in scientific research", in International Conference on Information, pages 303-313, https://arxiv.org/pdf/1802.03074.pdf/ (accessed on 4 March 2020).

Notes

[1] Knowledge-based capital comprises computerised information, innovative property and economic competencies (Corrado, Hulten and Sichel, 2004).

[2] www.data.gov.

[3] The total exceeds 86% because some categories overlap (e.g. a same repository can have partly open, partly restricted and partly closed datasets).

[4] Chapter 4 will address those concerns.

2 International initiatives promoting enhanced access to publicly funded data for science, technology and innovation

This chapter gives an overview of international initiatives in favour of enhanced access to publicly funded data for science, technology and innovation (STI), both on data from publicly funded research, as well as initiatives related to the broader category of public-sector information, which is increasingly being used as an input to STI. It starts by an overview of the achievements of the 2006 OECD *Recommendation of the Council concerning Access to Research Data from Public Funding*. It then reviews European Commission initiatives, as well as community-driven initiatives, and G7 and UNESCO work on the subject.

Initiatives pertaining to data from publicly funded research

The OECD Recommendation of the Council concerning Access to Research Data from Public Funding

The OECD *Recommendation of the Council concerning Access to Research Data from Public Funding* (hereafter "the Recommendation") and the related *OECD Principles and Guidelines for Access to Research Data from Public Funding* (OECD, 2006) resulted from work of the OECD Committee on Scientific and Technological Policy (CSTP) accomplished since 2001, including the Ministerial Declaration on Access to Research Data from Public Funding (OECD, 2004). They represent an important step in multilateral *efforts* to create the conditions for opening up access to data in the field of science, technology and innovation (STI).

A subsequent review by the OECD in 2009 showed many positive impacts of the Recommendation, notably the advancement of science through an accelerated research process, the emergence and development of new avenues of research beyond the initial context in which the data were collected and improved research quality (OECD, 2009). Other positive impacts included: i) enhancing research collaboration, both domestically and globally; ii) facilitating cross-disciplinary research; iii) preventing duplication of research; iv) conducting more research based on the same data; v) counteracting misconduct and increasing transparency; vi) validating and/or correcting previous results through re-analysis; vii) training new researchers by replicating studies; viii) boosting public confidence in research results; and ix) improving the evaluation and accountability of public funding. The negative impacts mainly concerned intellectual property right (IPR) issues and the costs associated with enabling access to data. The review highlighted issues such as access to data by foreign researchers and entities, researchers' desire to be the first to publish results, and the actual or perceived IPR problems involved in collaborative projects between the public and private sector. In addition, costs associated with documenting and describing data and collection procedures were mentioned, as well as the fact that the original data producer would have to bear these costs (OECD, 2009).[1]

In 2017, the OECD-CSTP conducted a survey on access-to-data policies among policymakers from 27 countries, particularly within science or assimilated (i.e. technology, innovation and education) ministries. A total of 55 institutions responded to the questionnaire; for most countries the response came from Ministries of Science or assimilated (Technology, Innovation, Education and Science, etc.). However, representatives of research institutes, funding agencies and repositories were also broadly represented (Figure 2.1).

Influence of the OECD Recommendation of the Council concerning Access to Research Data from Public Funding

Respondents to the 2017 OECD-CSTP survey were asked to assess the influence of the Recommendation on each of the initiatives undertaken (Figure 2.2).

Overall, the number of initiatives has grown over time, demonstrating the rising importance of data access. Even though some selection bias could exist – whereby respondents are more likely to include more recent initiatives – this trend seems quite pronounced, with only 6 initiatives in 2006-08, compared to 67 initiatives in 2015-17.

Examining the qualitative responses, the Recommendation is seen to have had a strong influence on the following types of initiatives:

- strategies, policies and laws focused on open access to data in research
- policies and strategies addressing the knowledge-based economy and society
- studies and guidelines on research-data management.

Figure 2.1. Respondent institutions

Institutions that responded to the 2017 OECD-CSTP survey concerning policy practice in supporting enhanced access to data for STI

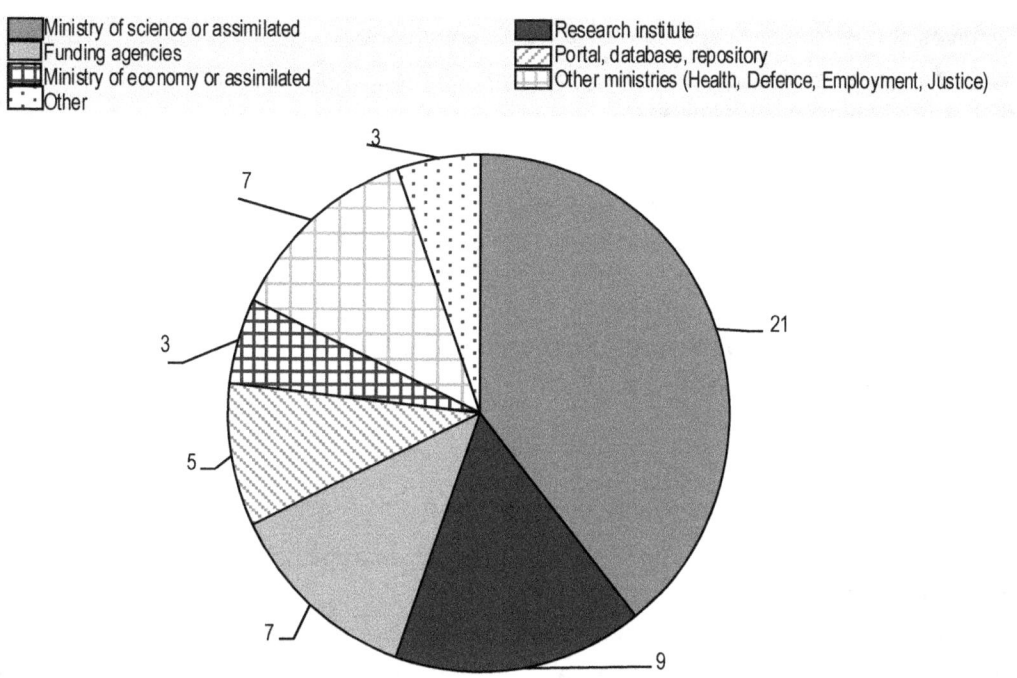

Source: Answers from OECD member countries and partner economies.

StatLink https://doi.org/10.1787/888934112538

Figure 2.2. Influence of the OECD *Recommendation of the Council concerning Access to Research Data from Public Funding*

From the 2017 OECD-CSTP survey concerning policy practice in supporting enhanced access to data for STI

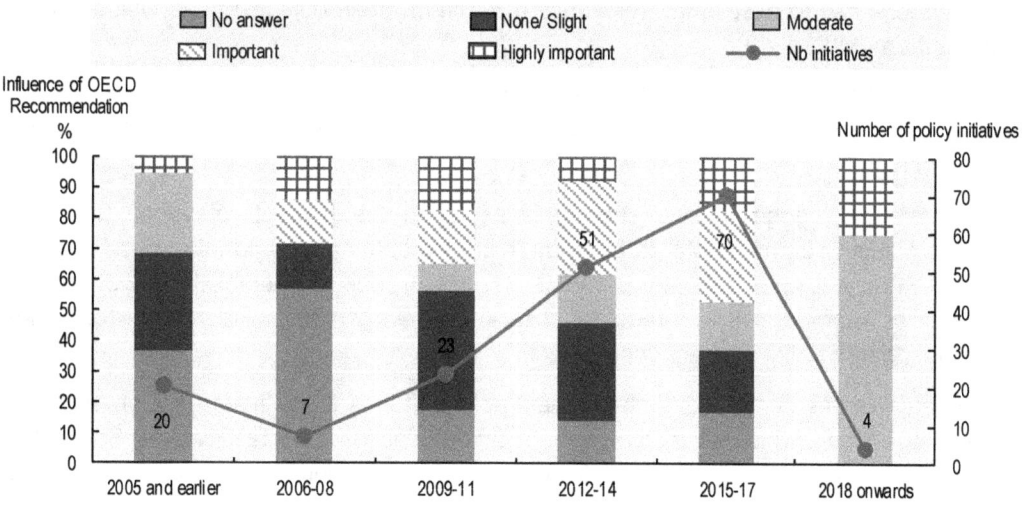

Note: A relatively stable proportion of answers state the Recommendation's influence on around 40% of all initiatives was "important" or "highly important".
Source: Answers from OECD member countries and partner economies.

StatLink https://doi.org/10.1787/888934112557

The Recommendation is likely to have had less impact in the following situations:

- broader open-government and public-sector information (PSI) initiatives, which cover data from public research as one of many different subsets of PSI (see Table 1.1), and as such do not focus on specificities of research data
- initiatives concerning data collected by private-sector entities
- initiatives in partner economies that have not yet adhered to the Recommendation (but are starting to consider it as a policy reference)
- in a few cases, lack of awareness of the Recommendation (even among OECD member countries).

Respondents also mentioned that although the Recommendation remains an important reference, they are also taking into account other international guidelines, such as the European Commission 2012 Recommendation on access to and preservation of scientific information (European Commission, 2012), the European Open Science Cloud (EOSC) (ZBW – Leibnitz Information Center for Economics, 24 January 2016), the Open Government Partnership (Open Government Partnership, 2014) and the Amsterdam call for open science (Government of the Netherlands, 2016).

Achievement of Recommendation objectives

The Recommendation's aims and objectives for advancing the data-access agenda include:

1. promoting a culture of openness and sharing of research data among the public-research communities within OECD member countries and beyond
2. stimulating the exchange of good practices in data access and sharing
3. saising awareness about the potential costs and benefits of restrictions, and limitations on access to and sharing of research data from public funding
4. highlighting the need to consider data access, and share regulations and practices, when developing OECD member countries' science policies and programmes
5. providing a commonly agreed framework of operational principles for establishing research-data access arrangements in OECD member countries
6. offering recommendations to member countries on improving the international research-data sharing and distribution environment.

Respondents were asked to assess the relevance of each of these objectives, as well as current policy effort related to them (Figure 2.3).

Ranking between 3.9 and 4.5 on the Likert scale, the overall relevance of the objectives is high, tending towards "very high" in some cases. Policy effort averages between 3 and 4 on the Likert scale, ranging from moderate to high.

Although Objective 2 on good-practice sharing is assigned the highest relevance, the policy effort dedicated to this objective is "moderate" rather than "high". Respondents report that good-practice sharing occurs mostly in seminars and conferences. Some countries report the establishment of communities of practice, as well as membership in international fora. However, many respondents report insufficient structured institutional or policy effort in this respect.

The objective perceived as commanding the highest policy effort is Objective 1 on promoting a culture of openness and data sharing, which is the core goal of most open access to data initiatives. Some respondents (Japan and Korea) point to differences across disciplines, with physics and biomedical sciences spearheading the trend towards openness.

Objective 4, including data and sharing in science policies, also ranks highly in both relevance and policy effort. Some countries, such as Finland, the Netherlands and France have implemented comprehensive open-science policies; others, such as Norway, include data access in individual policies, including in funding regulations. Moreover, Denmark, Korea and Canada report that open access to data is not always specifically addressed in *science* policies. Canada, for example, deals with access issues through the Federal Government policy instruments (the Policy on Service and Digital and the Directive on Open Government) and a tri-agency open-access policy on publications and statement of principles on data management.

Figure 2.3. Relevance and current policy effort dedicated to achieving the objectives of the OECD *Recommendation of the Council concerning Access to Research Data from Public Funding*

From the 2017 OECD-CSTP survey concerning policy practice in supporting enhanced access to data for STI

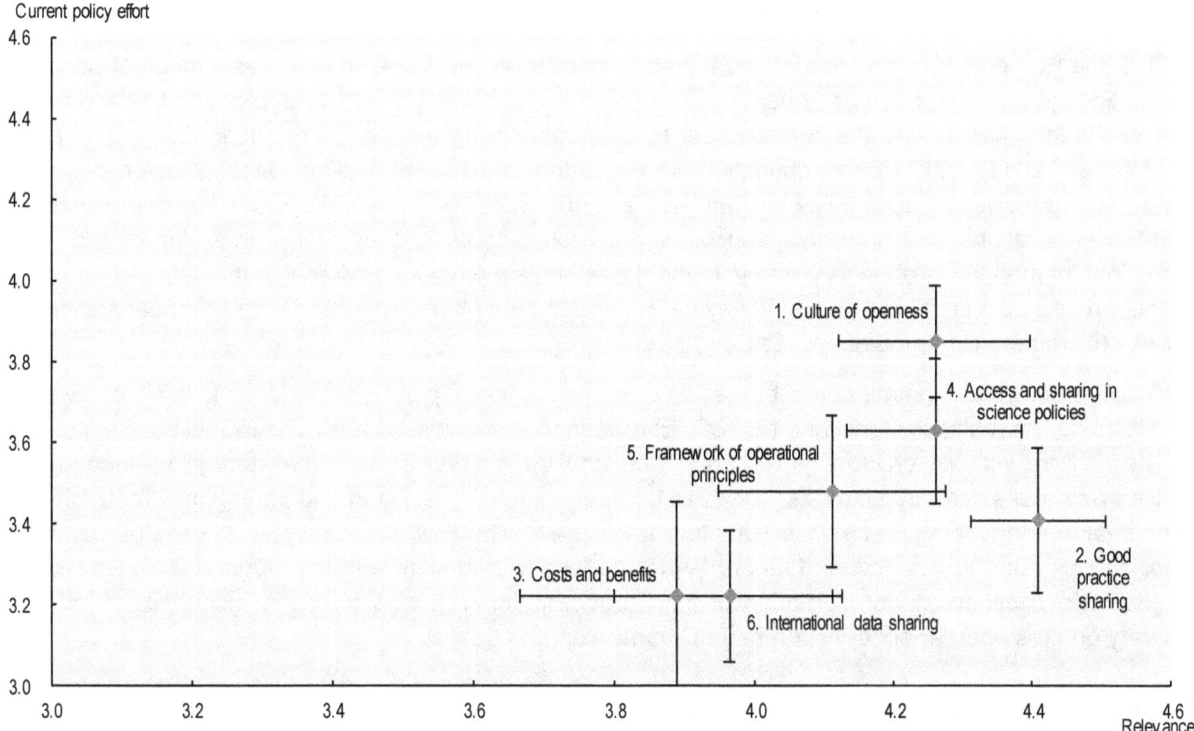

Notes: An average score was computed from the responses on a Likert scale: (1= "none"; 2= "slight"; 3= "moderate"; 4= "high"; 5= "very high"). The average assigns equal weight to countries, i.e. if several assessments are submitted by country, those were averaged out first, and the average score of the country was used for the overall average. The error bars show the statistical error on the mean score, at a 68% confidence level.
Source: Answers from OECD member countries and partner economies.

StatLink https://doi.org/10.1787/888934112576

The framework of operational principles (Objective 5) ranks as the next priority. Respondents quote the formulation of the findability, accessibility, interoperability and reusability (FAIR) principles (Wilkinson et al., 2016) as a good basis for the framework, but further efforts are needed to operationalise those principles. Canada has issued a specific guidance document on releasing scientific data and is addressing the licensing issue under an Open Government Licence. The European Commission has published the Guidelines on FAIR Data Management in Horizon 2020 (H2020) which encompasses rules on open access to scientific publications and open access to research data (European Commission, 2016a), and requires a data-management plan as a key element for good data management in research projects.

Awareness of the potential costs and benefits of limiting access to data and sharing (Objective 3) ranks relatively low compared to other objectives, especially considering the policy effort, which is seen as moderate. Although respondents agree the objective is important, the policies in place are relatively limited. The European Commission is preparing to conduct a study that will evaluate the costs of FAIR data to the EU economy, as well as estimate the costs and benefits of implementing the FAIR principles. Moreover, a number of pertinent research projects are being funded through H 2020.

Improving the international data-sharing and distribution environment also ranks relatively low. Countries report participating in many international fora, including the Research Data Alliance (RDA), the European Science Cloud and the GO (Global Open) FAIR initiative, the Group of Seven (G7) Working Group on Open Science, the Group of Senior Officials on Global Research Infrastructure and the Global Science Forum. However, the issue of data access is primarily addressed at the national level. The final statement from the Group of Twenty (G20) meeting in 2016 also "encourage[d] discussion on open science and access to publicly funded research results on findable, accessible, interoperable and re-usable (FAIR) principles in order to increase collaboration on science and research activities" (G20, 2016).

European Commission initiatives

In 2012, the European Commission issued a Recommendation on access to and preservation of scientific information calling for co-ordinated open access to scientific publications and data, preservation and reuse of scientific information, and the development of e-infrastructures among EU member states (European Commission, 2012). This Recommendation was updated in 2018, extending the application of "open access that policies aim to provide researchers and the public at large with access to peer-reviewed scientific publications, research data and other research outputs free of charge in an open and non-discriminatory manner as early as possible in the dissemination process, and enable the use and re-use of scientific research results". It also underlines data-management planning is becoming a standard scientific practice (European Commission, 2018a).

In 2016, the European Commission published its Open Innovation, Open Science, Open to the World vision, incorporating its ambitious plans for an EOSC (European Commission, 2016b). The EOSC aims to provide EU researchers with an environment offering free and open services for data storage, management, analysis and reuse across disciplines, achieved through connecting existing and emerging infrastructures, adding value and leveraging past infrastructure investment. The EOSC is expected to develop common specifications and tools to ensure data are FAIR and legally compliant with the General Data Protection Regulation and cybersecurity legislation of the European Union. It also foresees mechanisms for cost recovery on cross-border access (European Commission, 2018a).

Community-led initiatives

The FAIR data principles were developed by a diverse set of stakeholders representing academia, scholarly publishers, industry and funding agencies. They are now becoming a mainstream reference for policy makers (Wilkinson et al., 2016) (Table 2.1).

The RDA is an international forum which also enables the discussion and sharing of good practices. It was initiated in 2013 by the European Commission, the United States National Science Foundation and National Institute of Standards, and the Australian Government's Department of Innovation, with the goal of building the social and technical infrastructure to enable open sharing of data. As of September 2017, the RDA had 43 organisational members and 6 000 individual members from 130 countries. RDA operates working groups that work on recommendations, which can be adopted as standards, as well as a number of supporting outputs (RDA, 2014). RDA recommendations address a broad range of issues related to interoperability, data citation, data catalogues and work flows for research-data publishing (RDA, 2017).

The Committee on Data (CODATA) of the International Council for Science (ISC) was founded in 1966 to further the ISC vision of science as a global public good. CODATA promotes the FAIR principles and, more broadly, global collaboration to advance open science, and improve the availability and usability of data in all areas of research. The main priority areas of CODATA are: i) promoting principles, policies and practices for open data and open science; ii) advancing the frontiers of data science; and iii) building capacity for open science, by improving data skills and the functions of the national science systems necessary to support open data (CODATA, 2016).

Table 2.1. Overview of FAIR principles

FAIR principles	Action items	Technical requirements
Findable: data should be easily found by humans and machines alike	• Establish portals and open-science clouds	• Globally unique and persistent identifiers • Data indexed in a searchable database
Accessible: as open as possible, as closed as necessary	• Use open licensing whenever possible • Establish trusted-user access for more sensitive datasets	• Machine readability • Standardised communication protocol • Metadata are accessible – even after the data are no longer available
Interoperable: datasets need to be combinable with other datasets	• Three aspects of interoperability: semantic (taxonomy), legal (rights) and technical (machine readability) • Standard-setting activities	• Semantic interoperability – common vocabulary • Data include relevant references to other datasets
Reusable: it must be possible to reuse and further process data in future research projects	• Data curation • Open Archival Information System (OAIS)-compliant repositories	• Metadata are exhaustive • Data describe multiple precise and appropriate properties • Data are released with a clear and accessible data licence • Data are connected to their origin • Data meet standards relevant to the field

Source: OECD (2018), *OECD Science, Technology and Innovation Outlook 2018*, https://doi.org/10.1787/sti_in_outlook-2018-en.

Founded in 2011, Science Europe is an association of European research-funding organisations that represents the scientific community within the common European research area. Science Europe advocates free access to publications and scientific data, good practices in research-data management, and reform of copyright law to ensure that legally accessed content can be freely mined without additional permission and cost (Science Europe, 2015).

A recent initiative that is gaining momentum is the set of FAIR principles (Wilkinson et al., 2016). These principles are championed by FORCE11 – The Future of Research Communications and E-scholarship, a non-government, multidisciplinary community established after the 2011 "Beyond the PDF" meeting in San Diego and dedicated to transforming scholarly communications through technology. Academics and policymakers in many countries are quickly adopting FAIR principles as a new reference. Among recent analyses relative to FAIR, one can cite the report and action plan of the Expert Group set up by the European Commission "Turning FAIR Data to Reality", published in 2018 (European Commission, 2018b).

The Transparency and Openness Promotion (TOP) guidelines were created by journals, funders and societies to align scientific ideals with existing practices. The TOP guidelines and standards cover data citation; data transparency; software; research materials, design and analysis; preregistration of study and analysis plans; and replication. Journals select the transparency standards they wish to adopt and decide on a level of implementation for each standard (TOP, 2014).

Other initiatives

In 2011, UNESCO adopted a Revised Draft Strategy on UNESCO's Contribution to the Promotion of Open Data to Scientific Information and Research (UNESCO, 2011). The document calls for UNESCO to provide policy advice on the development of comprehensive national open-access to data policies, strengthen the capacities to adopt open access to data, serve as a clearing house and inform the global open-access to data debate. A year later, UNESCO published its Policy guidelines for the development and promotion of open access (Swan, 2012).

The Group of Seven (G7) Science and Technology Ministers' Meeting held in Tsukuba City (Japan) in 2016 decided to establish a working group on open science. The working group identified three action points as essential to the transition to open science: i) adopting a common vision; ii) adapting incentive and reward systems; and iii) developing a federated research-data infrastructure (ZBW – Leibniz Information

Center for Economics, 24 November 2016). Based on the report from the working group, the G7 Science Ministers identified incentives for researchers and infrastructures as new targets of the working group in Turin (Italy) in 2017. As mandated by the G7 Science Ministers, the working group continues discussing the promotion of global open science.

Initiatives related to public-sector information

Beyond data resulting from publicly funded research, a broader set of PSI is very relevant to research, as discussed in Chapter 1 and Figure 1.2. Therefore this section presents a brief overview of these initiatives.

The OECD *Recommendation of the Council for Enhanced Access and More Effective Use of Public Sector Information* (OECD, 2008) provides policy guidelines to improve access to and increase the use of PSI through greater transparency, enhanced competition and more competitive pricing. It aims to enhance the economic and social benefits derived from better access, and wider use and reuse of PSI. A subsequent evaluation concluded the Recommendation significantly contributed to policy making in this domain (OECD, 2015). It was used as a reference for developing practical approaches to PSI, e.g. to apply PSI principles in portal and licence design or strengthen some aspects of transparency and proactive openness. The principles on openness, access and transparent conditions for reuse, asset lists, copyright and pricing were deemed particularly useful (OECD, 2015).

In June 2013, the Group of Eight (G8) leaders signed the Open Data Charter, which sets five principles: i) open data by default; ii) data quality and quantity; iii) data usable by all; iv) releasing data for improved governance; and v) releasing data for innovation (G8, 2013).

The European Union followed up with an implementation plan featuring specific action points, such as committing to publish key datasets (including the budget of the European Union, EU Parliament election results, and data about EU staff, finance and contracts) under open licence; these are available on the European Union Open Data Portal, which now contains 11 600 datasets. The European Union also committed to encouraging all member states to apply the G8 Open Data Charter (European Commission, 2013).

The International Open Data Charter, initiated at the Open Government Partnership Global Summit in 2015, is a collaboration between governments and experts around six principles stipulating how governments should publish information: i) key datasets[2] are open by default; ii) data is timely and comprehensive; iii) data is accessible and usable (i.e. available online in machine-readable format, available in bulk for easy downloading, free of charge and open-licensed); iv) data is comparable and interoperable; v) data contributes to improved governance and citizen engagement; vi) data contributes to inclusive development and innovation. The International Open Data Charter has been joined by 17 national[3] and 30 local and subnational governments (International Open Data Charter, 2017).

Data integration – i.e. linking different datasets to fully exploit the significance of data – represents another major challenge for governments. In the United Kingdom, linking information from hospitals with the cancer-data repository and data from various screening programmes made it possible to recommend changes in medical protocols that should improve cancer survival rates. Data integration needs to be done in a way that preserves citizens' personal privacy. The Government of Australia recently launched an AUD 131 million data-integration partnership initiative to maximise the use and value of the government's data assets including to improve linkages among government datasets (Australian Department of the Prime Minister and Cabinet, n.d.).

References

Australian Department of the Prime Minister and Cabinet (n.d.), "Data Integration Partnership for Australia", https://pmc.gov.au/public-data/data-integration-partnership-australia (accessed on 20 January 2020).

CODATA (2016), "CODATA's mission", webpage, www.codata.org/about-codata/our-mission (accessed on 22 January 2020).

European Commission (2018a), "EOSC strategic implementation roadmap 2018-2020", Directorate-General for Research and Innovation, https://ec.europa.eu/research/openscience/pdf/eosc_strategic_implementation_roadmap_short.pdf#view=fit&pagemode=none (accessed on 3 July 2019).

European Commission (2018b), "Turning FAIR into reality", European Commission, Directorate-General for Research and Innovation, https://publications.europa.eu/en/publication-detail/-/publication/7769a148-f1f6-11e8-9982-01aa75ed71a1/language-en/format-PDF/source-80611283 (accessed on 20 January 2020).

European Commission (2016a), "H2020 Programme – Guidelines on FAIR data management in Horizon 2020", European Commission, Directorate-General for Research and Innovation, https://ec.europa.eu/research/participants/data/ref/h2020/grants_manual/hi/oa_pilot/h2020-hi-oa-data-mgt_en.pdf (accessed on 11 September 2019).

European Commission (2016b), "Open innovation, open science, open to the world – A vision for Europe", European Commission, Directorate-General for Research and Innovation, https://publications.europa.eu/en/publication-detail/-/publication/3213b335-1cbc-11e6-ba9a-01aa75ed71a1 (accessed on 28 February 2020).

European Commission (2013), *Digital single market: EU implementation of the G8 Open Data Charter*, European Commission, https://ec.europa.eu/digital-single-market/en/news/eu-implementation-g8-open-data-charter (accessed on 11 January 2020).

European Commission (2012), "Commission Recommendation of 17.7.2012 on access to and preservation of scientific information", European Commission, https://ec.europa.eu/research/science-society/document_library/pdf_06/recommendation-access-and-preservation-scientific-information_en.pdf (accessed on 12 September 2019).

G8 (2013), "G8 open data charter and technical annex", UK Cabinet Office policy paper, https://www.gov.uk/government/publications/open-data-charter/g8-open-data-charter-and-technical-annex (accessed on 28 February 2020).

G20 (2016), "Statement of the G20 science, technology and innovation Ministers meeting", https://www.mofa.go.jp/mofaj/files/000205642.pdf (accessed on 11 January 2020).

Government of the Netherlands (2016), "Amsterdam call for action on open science", document based on input from the April 2016 Amsterdam Conference "Open Science – From Vision to Action", https://www.government.nl/binaries/government/documents/reports/2016/04/04/amsterdam-call-for-action-on-open-science/amsterdam-call-for-action-on-open-science.pdf (accessed on 12 September 2019).

International Open Data Charter (2017), "Open up guide: Using open data to combat corruption", https://opendatacharter.net/open-guide-using-open-data-combat-corruption/ (accessed on 5 September 2019).

OECD (2018), *OECD Science, Technology and Innovation Outlook 2018*, OECD Publishing, Paris, https://doi.org/10.1787/sti_in_outlook-2018-en.

OECD (2015), "Assessing government initiatives on public sector information: A review of the OECD Council Recommendation", *OECD Digital Economy Papers*, No. 248, OECD Publishing, Paris, https://dx.doi.org/10.1787/5js04dr9l47j-en.

OECD (2009), "Access to research data: Progress on implementation of the Council Recommendation", 23-24 March, Paris, unpublished.

OECD (2008), *Recommendation of the Council for Enhanced Access and More Effective Use of Public Sector Information*, OECD, Paris, https://legalinstruments.oecd.org/en/instruments/OECD-LEGAL-0362.

OECD (2006), *Recommendation of the Council concerning Access to Research Data from Public Funding*, OECD, Paris, https://legalinstruments.oecd.org/en/instruments/OECD-LEGAL-0347 (accessed on 27 February 2020).

OECD (2004), *Declaration on Access to Research Data from Public Funding*, OECD, Paris, https://legalinstruments.oecd.org/en/instruments/OECD-LEGAL-0321 (accessed on 10 January 2020).

Open Government Partnership (2014), "Open government guide: All topics", webpage, Open Government Partnership, www.opengovguide.com/glossary (accessed on 14 January 2019).

RDA (2017), "All recommendations and outputs", webpage, Research Data Alliance, https://www.rd-alliance.org/recommendations-and-outputs/all-recommendations-and-outputs (accessed on 25 September 2019).

RDA (2014), "RDA governance document", Research Data Alliance, http://dx.doi.org/10.15497/RDA00001 (accessed on 24 February 2020).

Science Europe (2015), "Text and data mining and the need for a science-friendly EU copyright reform", briefing paper, www.scienceeurope.org/our-resources/briefing-paper-on-text-and-data-mining-and-the-need-for-a-science-friendly-eu-copyright-reform.

Swan, A. (2012), *Policy Guidelines for the Development and Promotion of Open Access*, UNESCO, Paris, https://unesdoc.unesco.org/ark:/48223/pf0000215863 (accessed on 28 February 2020).

TOP Guidelines Committee (2014), *Transparency and Openness Promotion (TOP) Guidelines*, Open Science Framework, Centre for Open Science, https://osf.io/ud578/?_ga=2.92744984.408225042.1583516285-1827136234.1583516285 (accessed on 9 March 2020).

UNESCO (2011), "Revised draft strategy on UNESCO's contribution to the promotion of open access to scientific information and research", programme and meeting document, UNESCO, 36th General Conference, https://unesdoc.unesco.org/ark:/48223/pf0000213342 (accessed on 24 February 2020).

Wilkinson et al. (2016), "The FAIR Guiding Principles for scientific data management and stewardship", *Sci Data* Vol. 3, No. 160018, http://dx.doi.org/10.1038/sdata.2016.18.

ZBW – Leibnitz Information Center for Economics (24 January 2016), "GO-FAIR: A member states-up strategy for the EOSC implementation", ZBW MediaTalk blog, https://www.zbw-mediatalk.eu/2017/01/go-fair-a-member-states-up-strategy-for-the-eosc-implementation/ (accessed on 12 September 2017).

ZBW – Leibnitz Information Center for Economics (24 November 2016), "The G7 Open Science Working Group action points – Speeding up open science?", ZBW MediaTalk blog, https://www.zbw-mediatalk.eu/en/2016/11/the-g7-open-science-working-group-action-points-speeding-up-open-science/ (accessed on 20 September 2017).

Notes

[1] It should be noted here that current understanding of business models for data provision will be further discussed in Chapter 4.

[2] Such as budget, spending, contracting, land ownership, company registries, legislation and election results.

[3] Australia, Argentina, Chile, Colombia, Costa Rica, France, Guatemala, Italy, Mexico, Panama, Paraguay, the Philippines, Sierra Leone, Korea, Ukraine, the United Kingdom and Uruguay. Only seven of these are OECD member countries.

3 Current policies in favour of enhanced access to data in OECD member countries and partner economies

This chapter presents an overview of current policies promoting enhanced access to data in OECD member countries and partner economies. It is based on the responses a survey implemented by the Committee for Science and Technology Policy of the OECD in 2017, the responses to the 2017 edition of the European Commission-OECD science, technology and innovation policy survey, as well as the case studies contributed by 18 countries during 2018.

Overview of policy initiatives promoting enhanced access to data and open science

At the national level, the 2017 European Commission (EC)-OECD science, technology and innovation (STI) policy survey (EC/OECD, 2018) asked OECD member countries and partner economies to provide information about policy initiatives supporting open science and open access. The 181 policy initiatives cited cover the following areas (Box 3.1):

- support for research infrastructure
- national policies and strategies in favour of open access to data (often linked to broader open-science strategies or open-government initiatives)
- creation of governance bodies to foster open access
- network and collaborative initiatives aiming to facilitate open access to data.

In the OECD-CSTP survey on access-to-data policies conducted in 2017, respondents from 27 countries listed a total of 171 policy initiatives. Roughly one-half of the initiatives concerned public research in general; one in six initiatives concerned a specific sector; and one-third were broader initiatives concerning public-sector information (PSI), including data from public research (Figure 3.1). Roughly one-half of the initiatives concern policy documents (including strategies) and legal measures.

To deepen the understanding of policy practice in the domain of enhanced access to data for STI, the OECD-CSTP contributed specific case studies of policies, sharing experiences and providing transferable learnings.

The case studies (OECD, 2018) cover a broad spectrum of initiatives, ranging from infrastructures and portals to national strategies. The approaches are diverse, with a mix of top-down and bottom-up initiatives (Table 3.1).

Figure 3.1. Scope of policy initiatives reported in the survey

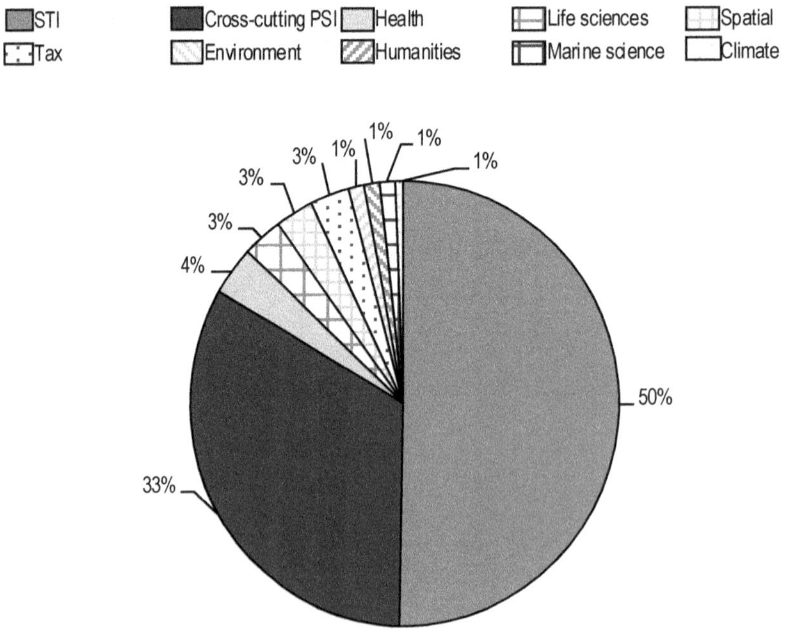

- STI: 50%
- Cross-cutting PSI: 33%
- Health: 4%
- Life sciences: 3%
- Spatial: 3%
- Tax: 3%
- Environment: 1%
- Humanities: 1%
- Marine science: 1%
- Climate: 1%

Note: PSI = public-sector information; STI = science, technology and innovation.
Source: Answers from OECD member countries and partner economies.

StatLink https://doi.org/10.1787/888934112595

Box 3.1. Instruments concerning data access from the 2017 EC/OECD STI policy survey

- The 74 reported initiatives (42%) supporting research infrastructures include portals offering open access to publications, repositories and archives for scientific data, search engines, virtual networks and clouds connecting individual physical repositories. Examples include the European Open Science Cloud (EOSC) and the Research Data Infrastructure for Open Science in Japan. In some cases (Australia, Estonia, Finland), open-data infrastructure is treated within a national strategy on research infrastructures.
- The 58 reported initiatives (34%) concerning national strategies and policies for open access to data and publications include:
  - Dedicated strategies and policies for open access to data and publications both at the policy-making level (Czech Republic, Korea, New Zealand, Norway, Slovenia and United Kingdom) and at the funding-agency level (Australia, Austria, Belgium-Federal, Canada, Lithuania, Nordic Council of Ministers, Netherlands, Norway, Portugal, Switzerland and United Kingdom); and a specific Memorandum in the United States from the White House Office of Science and Technology Policy (OSTP) instructing government funding agencies to increase access to data.
  - Open-data access within open-science policies (Chile, Colombia, Cyprus,[1] Denmark, Estonia and the Netherlands); the Open Innovation Strategy (Austria); the National Innovation and Science agenda (Australia); respective amendments in the Law on Scientific Activity are in progress (Latvia); and a specific Law 310/2014 for Public Research, which focuses on co-operation between business and academia (Greece).
  - In France, a specific "Law for a Digital Republic", followed by a National Plan for Open Science in France, promoting open access to scientific publications, optimal use and reuse of research data, and adapting evaluation and reward systems to bring them into line with the objectives of open science.
  - Open-data access, integrated within open-government and PSI initiatives (Australia, Argentina, Brazil, Canada, Sweden and United States).
  - Open access addressed within the law on official statistics (Norway).
  - Bottom-up approaches through institutions and projects (Centre national de la recherche scientifique and Institut national de la recherche agronomique in France; Infrastructures and Standards for Open Science in Spain; the University of Malta; universities in Slovenia; and the Concordat on Open Research Data in the United Kingdom).
- The reported 12 initiatives (7%) aiming to create or reform a governance body to foster open access include:
  - Etalab, a high-level, pan-governmental open-data platform in France that co-ordinates open-data and open-government initiatives, and is chaired by the national chief data officer, reporting to the Prime Minister of France
  - a national focal point (chief science advisor Canada, national chief data officer in France, point of reference in Slovenia) for access to and preservation of scientific data
  - an agency for information systems used in higher education and research (CERES – National Centre for Systems and Services for Research and Studies, Norway)
  - the Data Archiving and Network Services Institute, which facilitates data archiving and reuse, and provides training and consultancy (the Netherlands)
  - open-data institutes (Canada and the United Kingdom) supporting economic, environmental and social-value creation opportunities arising from open data.

- The 10 reported initiatives (4%) reported concerning networking and collaborative platforms to facilitate open access to data include:
  - OpenAIRE Advance, a network of repositories with 34 national open-science desks promoting open science as the default solution in Europe
  - library networks (HEAL Link in Greece, HAL and Persée in France)
  - open Access publishing platforms (J-Stage in Japan, OpenEdition in France)
  - the Datacite consortium, which enables researchers to attach a digital object identifier to research data
  - a data-analytics initiative linking disparate government datasets (Data61 in Australia)
  - co-operatives of research, educational and medical institutions (e.g. the Collaborative Organisation for ICT in Dutch Education and Research [SURF] co-operative in the Netherlands), aiming to promote innovation in information technology
  - a commercialisation marketplace (Open Data Exchange in Canada)
  - Support for digital facilities for the French research community (DORANIUM and Cat OPIDoR).
- The 5 initiatives (3%) undertaken based on formal consultations of stakeholder groups, including expert groups, include:
  - working groups and committees for open science and open access to scientific data (e.g. the European Commission Directorate-General for Research, Technology and Innovation; and initiatives in France, Greece, Ireland, Japan, Slovenia, Turkey and the United Kingdom)
  - an open-data forum advocating the development of open-data policies (United Kingdom).

1. Note by Turkey
The information in this document with reference to "Cyprus" relates to the southern part of the Island. There is no single authority representing both Turkish and Greek Cypriot people on the Island. Turkey recognises the Turkish Republic of Northern Cyprus (TRNC). Until a lasting and equitable solution is found within the context of the United Nations, Turkey shall preserve its position concerning the "Cyprus issue".
Note by all the European Union Member States of the OECD and the European Union
The Republic of Cyprus is recognised by all members of the United Nations with the exception of Turkey. The information in this document relates to the area under the effective control of the Government of the Republic of Cyprus.
Source: EC/OECD (2018), STIP Compass, https://stip.oecd.org/stip.html.

Table 3.1. Overview of initiatives covered by the case studies

	Top-down, driven by government or funding agencies	Bottom-up, driven by institutions
National strategies	• Korean Strategy to Promote Sharing and Use of Research Data for Innovative Growth • Norwegian National strategy on access to and sharing of research data • Finnish Open Science and Research Initiative (ATT) • Spanish law on STI	
Governance arrangements	• National chief data officer (France) • Infrastructures for Register-based Research (a government commission to the Swedish Research Council [SRC])	
Policies	• Slovenian National policy and Action plan of implementing of the national policy for open access	• UK Concordat on Open Research Data • Netherlands National Plan Open Science
Regulations	• EU Horizon 2020 (H2020) Open Research Data Pilot and Data Management Plan	• Data Management Plan Belgium Consortium
Infrastructure and portals	• Argentine Science and Technology Information Portal • Canada Open Government Portal • German National Research Data Infrastructure (NRDI) • Mexican Open Science Policy – Open Repositories Programme	• Colombian Biodiversity Information System (SiB Colombia) • DataFirst Open African Research Data Repository

Source: Case studies provided by OECD member countries and partner economies.

Motivation for policy initiatives promoting enhanced access to data for STI

Following up on initiatives to provide open access to publications, in which most countries have made significant progress, the focus is now shifting to enhancing access to data through open-data arrangements, which is generally less developed. Although more than 92% of universities in Europe have open-access policies for publications, or plan to do have them in the near future, less than 28% had guidelines in place for open access to data. The main institutional barriers to promoting research-data management and/or open access to research data are: different "scientific cultures" within the university; the absence of national guidelines or policies; limited awareness of benefits; legal concerns; and technical complexity (Morais and Borrell-Damian, 2018).

An EC survey found that by 2016, only 8 out of 31 European countries surveyed had adopted a national policy or strategy encouraging or mandating open access to research data; of these, only 1 has actually been implemented. A further 14 countries are discussing such a strategy or policy, and another nine countries are not considering such policies (European Commission, 2018a). The same study concludes that funders in a majority of countries (17 out of 31) are now including data-management plans as a requirement, or even as evaluation criteria for applicants.

Analytically speaking, and as discussed in Chapter 1, there exist several rationales in favour of enhanced access to data for STI. These include: i) exploiting opportunities to create new scientific insights; ii) promoting innovation and economic growth; iii) enhancing social welfare for individuals and society at large; iv) enhancing the reproducibility of scientific results; v) supporting meta-analyses; vi) facilitating education; vii) avoiding duplication of research; and viii) improving governance.

In practice, the motivations vary according to the specific context and international influences.

The Korean, Norwegian, and Slovenian cases describe national strategies on access to and sharing of research data. In Norway, the strategy reinforced previous policy principles that were "open by default" for publicly funded research – particularly research financed by the Research Council of Norway – driven by the finding that the quantity of data available was below expectations for an "open by default" setting. Moreover, reuse of said data did not follow best practices (Norwegian Ministry of Education and Research, 2018). In Slovenia, the goal is even more ambitious, and involves suppression of subscription and copyright limitations for access to and reuse of scientific information.[1] The Slovenian policy invokes the implementation of the Hague Declaration on Knowledge Discovery in the Digital Age (Tramte, 2018).

In Korea, the prime motivation of the Strategy to Promote Sharing and Use of Research Data for Innovative Growth is also to enhance access, with the expectation of a strong increase in demand for research data that should feed data-driven innovation and enhance national competitiveness (Shin, 2018). To promote sharing and use of research data, the strategy proposes to achieve at least five policy aims over the next five years: i) support the establishment of a research-data centre in each field of research and help it grow within the research community; ii) establish a national research-data platform connecting field-oriented data centres to one another; iii) provide education and training for data scientists/engineers to enhance their data skills and expertise; iv) develop a legal basis for sharing and use of research data, and adoption of data-management plans; and v) promote innovation and commercialisation activities inspired from shared research data (Steering Committee of the National Science and Technology Council, 2018). Since 2016, research outputs published in scientific journals resulting from research funded by more than 50% of public funds can be provided by their author as open access, following an embargo period ranging from 6 to 12 months –an exclusivity period granted to the author in order to be able to be the first to publish results based on the date she collected. In any case, the publisher must let the data published within these articles freely reusable after their publication. Such data also must be freely accessible for text and data mining (TDM).

France addresses access to research data in a broader context of "public and research data". The objective is to bring the data infrastructure to a state where it provides the public with effortless access to updated and reliable data. The aim is to render the inner works "invisible" to the public, similarly to the provision of electricity and fresh water. Publication of government and other public entities' data is mandatory. Opening of data also becomes mandatory for private data that considered of public interest under French law, e.g. data concerning the provision of public services such as energy and water, real-estate transactions or waste recycling (French Ministry of Economy, Industry and the Digital Sector, 2015). In order to abolish nontariff barriers to access (such as technical, legal, cultural or awareness issues), France has introduced the position of national chief data officer, who can also rely on the "Etalab" structure, to carry out his missions of co-ordinating government efforts to facilitate the provision, governance, production, circulation and reuse of government data, including research data (French Ministry of Higher Education, Research and Innovation, French Ministry of Higher Education, Research and Innovation , 2018).

Many of the initiatives reported relate to open science; this is a strong systemic tendency in science and technology policies of an increasing number of countries, including Finland, the Netherlands, France, Mexico and the European Union. The Netherlands and Finland contributed case studies describing their national plan for open science. France described the national chief data officer, who helps operationalise open government and open science. The European Union (European Commission, 2018a) and Mexico (CONACYT, 2018) presented specific initiatives supporting open-science policies. One of the pillars of open science is open access to data (and the e-infrastructures that enable it), alongside open research agenda-setting, open funding mechanisms, open access to publications, citizen science, open research infrastructures, open peer review and other open-science tools.

Finland was one of the pioneers of this movement, setting up a Finnish Research Data Initiative to advance open access to research publications, open access to research data and open research methods. The importance of these focus areas was confirmed in the opinions gathered from major stakeholders. In 2014, building on the previous projects, the Finnish Ministry of Education and Culture launched its latest and most holistic initiative, the Open Science and Research Initiative (ATT) (Haapamaki, 2018). Its aim was to incorporate open science and research in the full research process, in order to improve the visibility and impact of science and research in the innovation system and society at large. To guide the Finnish research system towards higher competitiveness and quality, the ATT advocates promoting a transparent, collaborative and inspirational research process, with measures that foster open publications, open research data, and open research methods and tools, as well as increasing skills and improving knowledge and support services in the open-science domain.

Motivation for open science itself can be found within the Amsterdam call for open science (Dutch Presidency of the Council of the European Union, 2016) – it includes increased openness, rapid, convenient and high-quality scientific communication between researchers and society at large, fostering better response to societal challenges, and providing business opportunities through the development of innovative products and services (Ministry of Education, Culture and Science, 2018).

The European Commission gives the following rationale for its H 2020 Open Research Data Pilot: "Making research data available benefits science by increasing the (re-)use of data and by improving transparency and accountability. Access to research data increases returns from public investment, reinforces open scientific inquiry, and enhances the quality and efficiency of scientific research and innovation, thus providing better business opportunity." It further underlines that opening up research data has the potential not only to improve scientific research and involve society, but also to contribute significantly to economic growth (through open innovation) (European Commission, 2018a).

These benefits concern notably innovative start-ups, e.g. in the context of the "app economy", as well as the use of data in the context of the reindustrialisation of Europe ("Industry 4.0"). Two key agendas, Open Science, Open Innovation, and Open to the world (European Commission, 2016a) and Digital Single Market support this vision through several strategic initiatives outlined in the EC Communication, "A European Cloud Initiative – Building a competitive data and knowledge economy in Europe" (European Commission, 2016b). The Open Science Agenda defines the open-access ambitions as follows: i) by 2020, all peer-reviewed scientific publications are freely accessible; and ii) by 2020, Findable, Accessible, Interoperable and Reusable (FAIR) data sharing is the default for scientific research.

The UK Concordat (UK Research and Innovation, 2016) has similar motivations: "The benefits from opening up research data for scrutiny and reuse are potentially very significant; including economic growth, increased resource efficiency, securing public support for research funding and increasing public trust in research. However, the Concordat recognises that access may need to be managed in order to maintain confidentiality, protect individuals' privacy, respect consent terms, as well as managing security or other risks." The Concordat's stated objective is to "ensure that the research data gathered and generated by members of the UK research community is made openly available for use by others wherever possible in a manner consistent with relevant legal, ethical, disciplinary and regulatory frameworks and norms, and with due regard to the costs involved" (Bruce, 2018).

Box 3.2. United States federal data strategy

The United States is developing a federal data strategy to provide a co-ordinated and integrated approach to using data to deliver on mission, serve the public and steward resources, while respecting privacy and confidentiality. The strategy incorporates four areas of exploration:

- Enterprise-data governance. This area refers to setting priorities for managing government data as a strategic asset, including establishing data policies; specifying roles and responsibilities for data privacy, security, and confidentiality protection; and monitoring compliance with standards and policies throughout the information lifecycle.

- Access, use and augmentation. This section is concerned with developing policies and procedures that enable stakeholders to effectively and efficiently access and use data assets. The strategy aims to increase these features by: i) making data available more quickly and in more useful formats; ii) maximising the amount of non-sensitive data shared with the public; and iii) leveraging new technologies and best practices to increase access to sensitive or restricted data, while protecting privacy, security and confidentiality, as well as the interests of data providers.

- Decision-making and accountability. This refers to improving the use of data assets by the federal government for decision-making and accountability, including both internal and external uses. This includes: i) providing high-quality and timely information to inform evidence-based decision-making and learning; ii) facilitating external research on the effectiveness of government programmes and policies that will inform future policy making; and iii) fostering public accountability and transparency by providing accurate and timely spending information, performance metrics and other administrative data.

- Commercialisation, innovation and public use. The final section relates to facilitating the use of federal government data assets by external stakeholders at the forefront of making government data accessible and useful through commercial ventures, innovation, or other public uses. This includes use by the private sector, scientific and research communities, and state and local governments for public-policy purposes, as well as for education and enabling civic engagement. Enabling external users to access and use government data for commercial and other public purposes spurs innovative technological solutions, and fills gaps in government capacity and knowledge. Supporting the production and dissemination of comprehensive, accurate and objective statistics on the state of the nation helps businesses and markets operate more efficiently.

Source: US Government (2019), "Public access to Federally funded research in the United States", https://community.oecd.org/servlet/JiveServlet/downloadBody/149047-102-1-263336/USA%20ANNEX%20to%20ENHANCED%20ACCESS%20TO%20PUBLICLY%20FUNDED%20DATA%20FOR%20SCIENCE_Final4secretariat.pdf.

Although the United States has not adopted the open-science framework, it recognises that openly accessible data contribute to scientific progress, catalyse innovation, and further international co-operation in science and technology to address global challenges (US Government, 2019). The broad availability of scientific information and underlying data allow for the critical review, replication and verification of findings that are central to the scientific method. Additionally, facilitating the free flow of information that underpins federal agency decision-making and advances in regulatory science – to the extent permitted by law – enables the public, the United States Congress, the media and industry to better understand the scientific basis for US federal regulatory decision-making. As part of this, the United States is developing a federal data strategy (Box 3.2.).

In Spain, the main motivation for including open access in the 2011 Law on STI was to comply with the relevant EU directives and to reinforce the position of the 42 open access institutional repositories that existed at that moment. However, the law did not explicitly mention data and was implicitly targeting publications. In a recent development, Spain wants to include research data in open-access policies and design a national strategy for open science. Since June 2018, the Spanish government agency FECYT has been undertaking the Infrastructures and Standards for Open Science pilot project with three public research organisations and their dedicated repositories (FECYT, 2019).

A specific issue relates to providing access to sensitive datasets. As noted in Chapter 1, enhanced access to data can be staged to different degrees, according to the community of stakeholders involved. The more sensitive the data, the more difficult it will be to open them to the general public, taking the risk of privacy breaches and malevolent uses. Therefore, different degrees of openness may include: i) open access with open licence; ii) public access with a specific licence that limits use: iii) group-based access through authentication; and iv) named access explicitly assigned by contract (OECD, 2019).

One example of such controlled access is the Secure Research Service provided by the United Kingdom's Office of National Statistics, which provides access to sensitive datasets to certified researchers (Office of National Statistics UK, n.d.). The French Law for a Digital Republic also foresees safe access to confidential data for approved researchers (French Ministry of Economy, Industry and the Digital Sector, 2015). Sweden has launched an initiative called "Infrastructures for Register-based Research" to address one of the most challenging issues – providing cross-disciplinary access to register data, which cover government registers, clinical-quality registers, biobanks and sample collections, and research databases (Eriksson and Nilsson, 2018). South Africa also reports a Secure Research Data Centre (based on the UK experience), comprising a "safe room" at the University of Cape Town (UCT) where disaggregated public data can be analysed by vetted researchers for purposes of cutting-edge research (Woolfrey, 2018).

Belgium has a specific initiative to support data-management planning. In a country that has yet to develop a holistic approach to enhanced access to data, such a pragmatic initiative facilitates compliance with EU regulations, particularly the data-management planning requirement for H 2020 projects (Laureys, 2018).

The Argentine, Canadian, Colombian, Mexican and South African case studies describe specific infrastructures, such as the Argentine Science and Technology Information Portal (Luchilo, D'Onofrio and Tignino, 2018), the Canadian Open Government Portal (Treasury Board of Canada Secretariat, Open Government Team, 2018), the Colombian Biodiversity Information System (Escobar, Hernández and Agudelo, 2018) and the Mexican Open Institutional Repositories Programme. Their motivation follows more general policies and legal frameworks.

In Argentina, the portal was a logical consequence of a comprehensive normative context consisting of several decrees and laws, notably Law 27.275 on Access to Public Information and the corresponding Decree 117/16 on the creation of Data Opening Plan, and Law 25.467 on the Creation of Digital Repositories (Luchilo, D'Onofrio and Tignino, 2018).

In Canada, the Open Government Portal was established following public consultations on the digital-economy strategy and open government, which identified the need to make open data available in more usable and accessible formats. The motivation for establishing the portal in 2011 was to provide a single access point for using and sharing structured, machine-readable data and information from across the government, under an open and unrestrictive licence (Treasury Board of Canada Secretariat, Open Government Team, 2018).

The Colombian Biodiversity Information System (SiB Colombia) was developed to fulfil the country's obligation under the Convention on Biological Diversity, as part of the 1992 United Nations Conference on Environment and Development held in Rio de Janeiro, Brazil (better known as the "Rio Earth Summit"). This national infrastructure was established in early 2000 to facilitate free and open access to biodiversity data (Escobar, Hernández and Agudelo, 2018).

Mexico created its Open Institutional Repositories programme as part of the implementation of its national open-science policy (CONACYT, 2018).

In South Africa, DataFirst was primarily created as a full-service repository to actively promote long-term preservation, sharing and analysis of data from publicly funded research. The university-based repository was created on the premise that: i) proximity to researchers keeps the repository responsive to research-data needs; ii) such a service is seen as independent, largely free from political influence and not susceptible to funding changes as political regimes change; and iii) such a repository can draw on universities' skill base, such as a constant supply of student interns to work on data cleaning. This is an important requirement in the skill-constrained environments of lower and middle-income countries: a university base can ensure that data repositories are able to maintain a high level of service in the long term (Woolfrey, 2018).

Germany's National Research Data Infrastructure was motivated by the need to ensure better co-ordination of research-data management in the country. The National Research Data Infrastructure was established as an agreement between the federal government and the federal states (*Länder*) to: develop a co-ordinated, networked information infrastructure for the development of sustainable interoperable research-data management; establish accepted processes and procedures to standardise handling of research data in the scientific disciplines; create a reliable and sustainable range of services covering the overarching and subject-specific needs of research-data management in Germany; develop cross-disciplinary metadata standards for the comprehensive (re)usability of research data; connect German research-data infrastructures to European and international platforms; optimise the reusability of research data already collected, as well as the infrastructures in which they are embedded, thereby generating additional knowledge without the high costs of new data collection; and create a common basis for data protection and data sovereignty, integrity, security and quality (Philipps and Bodmann, 2018).

Initiation of policy initiatives and process

As shown in Table 3.1, the case studies cover a variety of approaches. The majority of initiatives (12 out of 17) are driven from the top (by ministries or funding agencies); four initiatives are driven by research consortia and/or higher education institutions.

Whatever the approach, achieving consensus on sharing research data is a challenging task. The top-down initiatives are far from command-and-control arrangements: stakeholder participation and consultation also play a large role. The following sections present those experiences.

Top-down initiatives driven by government and funding agencies

This group of top-down initiatives presents considerable variety. For one, they are initiated at different levels, beginning with the prime minister (in France), followed by the Ministry of Science and Technology or equivalent (in Argentina, European Union, Finland, Germany,[2] Korea, Norway, Slovenia, Spain)[3] and finally the funding-agency level (in Mexico and Sweden).[4]

The governance mode also varies considerably:

- France presented an initiative on the governance of open access to government data (including research data). This novel concept of National Chief Data Officer (NCDO) has a central role as a champion and co-ordination point for open data (French Ministry of Higher Education, Research and Innovation, 2018).

- In Germany, the Joint Science Conference of the federal states (Länder) (German federal government and German Bundesländer, 2018) and the federal government of Germany prepared an agreement signed by the federal government and the Länder establishing the National Research Data Infrastructure (Philipps and Bodmann, 2018).
- In Korea and Slovenia, the initiatives, although driven by the corresponding ministry, include active stakeholder participation throughout the process in dedicated working groups used for consensus-building about the policies. In Korea, a task force was set up to develop the national strategy, including range of stakeholders such as researchers and data scientists from universities and government research institutes, private-sector managers, policy experts and lawyers, organised into six working groups (Shin, 2018); in Slovenia beneficiaries of the national public-research funding were invited to participate in public consultation on the draft national open access strategy, and received 24 written comments (Tramte, 2018). Likewise, the Argentine science and technology information portal initiative has undertaken a phase of political consultations with key stakeholders to achieve consensus early in the process (Luchilo, D'Onofrio and Tignino, 2018).
- In the European Union, a broad stakeholder consultation preceded and was used as essential input for the initiative (European Commission, 2018a).
- Norway drafted its strategy based on analytical expert work within an interministerial working group; research stakeholders were not directly involved in the process (Norwegian Ministry of Education and Research, 2018).
- Mexico presents a three-level governance approach: the federal law, general guidelines and technical guidelines. Specific institutions are responsible for each level: the Congress of the Union is the only institution capable of modifying the federal law; the internal governance entities of the National Council of Science and Technology (CONACYT) can modify the second and third-level documents. Stakeholder input is only solicited for the second and third levels (CONACYT, 2018).

The following section synthesises the individual governance and process arrangements in the 12 country cases.

France: NCDO

The French initiative (French Ministry of Higher Education, Research and Innovation, 2018) has been set at the highest level, whereby the NCDO reports to the secretary of state for digital society within the prime minister's office. The mandate of the NCDO includes:

- co-ordinating the production, storage, diffusion and (re)use of data by the government administration
- organising and ensuring the wider diffusion of data (re)use, particularly for the purpose of public-policy evaluation, improvement and transparency of public action, and stimulation of research and innovation, while ensuring the protection of personal data and secrets safeguarded by the law
- proposing new policies to the prime minister, including, where appropriate, legislative or regulatory developments.

In consultation with the concerned administrations, the NCDO:

- proposes to the prime minister strategies (which may also involve innovative companies) for (re)use of the data produced, received or collected by the administrations as part of their public-service missions
- develops tools, benchmarks and methodologies for better data (re)use and greater use of data science within the government
- recommends technical solutions to increase the interoperability of information systems and data, including semantic data modelling
- conducts experiments on the use of data to enhance the effectiveness of public policies, contribute to the sound management of public funds and improve the quality of services provided to users.

The NCDO may also be questioned by any person regarding the flow of data.

In parallel, France is developing an open-science ecosystem as part of the Open Government Partnership roadmap. A key element is the establishment of an open-science committee, whose mandate will be to co-ordinate and support the transition towards open science in France. The committee will comprise four colleges: i) college for publications; ii) college for research data; iii) college for skill and training; and iv) college Europe and international. The committee's membership strives to achieve broad representativeness across research institutions and disciplines, including data science (BSN, 2018). Other elements of the planned ecosystem include specific operational items, such as the introduction of mandatory data-management plans, further development of the Scan search engine, development of "data papers" and monitoring schemes (Etalab, 2018).

Finland: Open Science and Research Initiative

Prior to the launch of the ATT, a working group on openness of research data was appointed to carry out preparatory work and identify major focus areas – including open data, which had already been highlighted in an earlier government strategy document. Meanwhile, the open-science movement was gaining ground internationally, and open science was a focus area in the recommendations of the European Commission and OECD. The working group worked on the themes for the Open Science Initiative against this background (Haapamaki, 2018).

The ATT initiative was launched by the Finnish Ministry of Education and Culture. It included a strategy group, chaired by the director in charge of science policy and comprising high-level representatives from universities, universities of applied sciences, research institutes, other ministries and funding bodies; and an expert group under the aegis of the strategy group, chaired by the secretary general of the initiative. A forum open to all interested parties convened annually to discuss current topics. More focused thematic roundtable discussions were also held within more targeted groups. All major stakeholders from the higher education and research sectors, as well as scientific publishing, were involved in the process.

The central point in the governance structure involved ensuring, on the one hand, stakeholder representativeness and on the other hand, flexibility to adapt to the changing nature of the open-access phenomenon. As a major funder, the Ministry of Education and Culture had the final say in decision-making.

Korea: Strategy to promote sharing and use of research data for innovative growth

The Korean strategy was initiated under the auspices of the Ministry of Science and ICT (MSIT). A task force was created, comprising an overarching team and five specialised subgroups. A broad range of stakeholders participated, including MSIT personnel, researchers and data scientists from higher education and public research organisations, private-sector managers, policy experts and lawyers. Following 26 meetings of the task force and a public hearing, the strategy was finalised and submitted to the National Science and Technology Council, a high-level interministerial co-ordination committee charged with developing a comprehensive national plan on STI and making adjustments to the research and development (R&D) programmes and budgets proposed by each ministry (Shin, 2018).

The implementation of the strategy is governed by the MSIT through the development and implementation of action plans. The National Research Foundation will develop its own policies in line with the strategy, and the National Research Council of Science and Technology[5] is also developing guidelines to assist government research institutes with managing their research data. The implementation began with pilot projects by government research centres that adopted data-sharing systems years ago and have created meaningful research outcomes. The pilot projects strive to improve existing mainstream data-sharing initiatives, specifically through: i) improving cloud analytics for genome data; ii) promoting the use of artificial intelligence (AI) on chemical-library data for new-drug discovery; iii) data sharing and research collaboration through setting up a web-based data platform on thermo-electronic materials and developing machine learning analytics; (iv) establishing data-sharing systems (R&D data bank) in the field of catalyst research; v) research-data retrieval at large research facilities with high-performance computing systems for research in integrative structural biology; and vi) expanding the scope of the research data collected and collaborating with government research institutes that produce useful data for AI development.

Norway: National strategy on access to and sharing of research data

Norway's strategy process was initiated and led by the Ministry of Education and Research, based on the 2016 white paper "Digital Agenda for Norway" (Ministry of Local Government and Modernisation, 2016), which presented the government's intention to develop strategies aiming to increase the accessibility of public data in five areas: culture, geodata, public expenditure, transport and research. It was also based on several international reports relevant to access to research data, as well as a survey of current practice in Norway led by the Research Council of Norway (DAMVAD Analytics A/S, 2014). An interministerial working group involving representatives of six other ministries oversaw the work. The implementation of the strategy will proceed through specific action plans for government and relevant agencies. Public research institutions will also be expected to: develop procedures for approving or deciding on their research projects' data-management plans; improve employees' skills (including researchers, data stewards and administrators); co-ordinate the development of new educational programmes; and encourage institutions and researchers to help develop relevant standards (Norwegian Ministry of Education and Research, 2018).

Slovenia: Research Data Management and Openness in Slovenia

The flagship policy initiative Research Data Management and Openness in Slovenia is led by the Ministry of Education, Science and Sport; its content is supported by Slovenian open-access experts (Tramte, 2018).

Drafting of the new Research and Development Activity Act began in 2015 within a working group comprising ministry staff and relevant stakeholders (i.e. beneficiaries of national public-research funding). Time was needed to achieve consensus among stakeholders, and the final public consultation took place in 2018. Basic regulatory-impact analysis was applied in the process and is part of the document to be submitted to both the Government and the National Assembly of the Republic of Slovenia.

The national pilot programme Open Access to Research Data is included in the National Strategy of Open Access to Scientific Publications and Research Data in Slovenia 2015-2020. Given the very modest Slovenian experience with open access to research data, the pilot programme was based on OECD and EC recommendations and international experience. Its results will be used to design the policy on open research data for the next period (after 2020).

The provisions of both the new Research and Development Activity Act and the pilot programme will be implemented primarily by the Slovenian Research Agency, Slovenia's main research-funding organisation.

Spain: Open-access dissemination in the Spanish Law on STI (Article 37)

The Ministry of Science, Innovation and Universities drafted the Law on STI in 2011 – notably Article 37, which prescribes open-access dissemination of scientific results. However, due to difficulties in its implementation, the government agency FECYT created a working group comprising national experts and representatives of institutions possessing their own open-access strategy to put forward a set of recommendations for the implementation of Article 37. This practical guide includes specific recommendations for R&D public funding agencies, universities and research centres, researchers and institutional subscribers to scientific journals. Among those recommendations was the creation of a monitoring commission for the implementation of Article 37 on open-access dissemination.

In a recent development, the State Plan for Research, Development and Innovation 2017-2020 (Spanish Ministry of Economy, Industry and Competitiveness, 2017) features important advances on open access to scientific publications and research data. Moreover, since June 2018, FECYT has been implementing the Infrastructures and Standards for Open Science pilot project, in collaboration with three public research organisations. The main goal is to promote the national infrastructures and the standards for information exchange needed to optimise the implementation of open-access policies; include research data in open-access policies; and help the Ministry of Science, Innovation and Universities in the designing of a national strategy for open science.

Germany: National Research Data Infrastructure

The National Research Data Infrastructure is supposed to systematically develop a comprehensive research-data management system, including standardised data management according to FAIR principles. It aims to sustainably secure and utilise research data, as a digital, regionally distributed and networked knowledge repository (Philipps and Bodmann, 2018).

It is an emerging initiative, established by an agreement between the federal government and the *Länder* in November 2018, funded by the Federal Ministry of Education and Research and the *Länder* (German federal government and German Bundesländer, 2018): the Joint Science Conference, which makes all fundamental financial decisions regarding the National Research Data Infrastructure, and also decides on its legal form and organisational set-up.

The National Research Data Infrastructure organisational structure consists of consortia, a consortium assembly, a scientific senate and a directorate.

Consortia will be composed of several existing institutions, which may include state and state-recognised universities, non-university research institutions, non-university research institutions, academies and other publicly funded information-related infrastructure facilities.

The consortium assembly shall consist of the elected spokespersons of each consortium. It determines the content-related and technical principles for the overall work of the consortia.

The scientific senate is the strategic body responsible for the overall strategic orientation of the National Research Data Infrastructure. It decides on standards across the consortia, as well as metadata standards and formats, on the basis of a proposal by the consortium assembly. The scientific senate also advises on the progress of the consortia's projects, and decides on the inclusion and integration of comprehensive services in the National Research Data Infrastructure.

Canada: Open Government Portal

Canada's Open Government Portal was initiated and is maintained by the Treasury Board of Canada Secretariat in the Chief Information Officer Branch (now Office of the Chief Information Officer). Federal departments and agencies are responsible for maximising the release of government information and data of business value; each of the federal government's Science-based Departments and Agencies releases open-data and/or information resources through the official portal.[6] Additionally, the Government of Canada recently established the position of Chief Science Advisor of Canada, who works to ensure that government science is fully available to the public and that government scientists can speak freely about their work (Treasury Board of Canada Secretariat, Open Government Team, 2018). As part of Canada's 2018-2020 Plan, there is a commitment to "Promote open science and actively solicit feedback from stakeholders and federal scientists on their needs with respect to open data and open science."

The portal provides access to a broad range of scientific data and information resources, including more than 60 000 geospatial datasets,[7] 6 000 data and information resources on the subject of "science and technology", data and information from science-based departments and agencies such as Environment and Climate Change Canada, Natural Resources Canada, Health Canada, and National Research Council Canada.

Canada's original open-government portal was created following consultation with Canadians on a digital-economy strategy and open government. The results stressed the importance of providing open access to PSI and data, in particularly the need to improve the availability of data for researchers and the private sector, with fewer restrictions on reuse of these information assets. Regular consultations include an online dialogue to generate potential ideas for the plan, a series of in-person workshops in cities across the country, a series of thematic webinars, an online questionnaire on the portal and social media presence, including tweets and blog posts about consultation opportunities.

Argentina: The Argentine Science and Technology Information Portal

The Science and Technology Information Portal was created on the basis of a comprehensive pre-existing legal and regulatory framework related to open access to the scientific and technological production and to the primary data of the scientific investigations carried out with funds from the government. This includes the Law on the Creation of the National Science and Technology system, the Law on the Creation of Digital Repositories, the Law on Access to Public Information, and appropriate decrees. Nevertheless, stakeholders expressed reluctance about providing access to data (Luchilo, D'Onofrio and Tignino, 2018).

The lead department for the portal is the Secretariat of Scientific and Technological Articulation of the Ministry of Science, Technology and Productive Innovation. Consensus was built with the science and technology institutions within the Inter-institutional Council of Science and Technology, which structures the national STI system. Under the co-ordination of the National Directorate of Programmes and Projects, meetings convened participating stakeholders. Individual meetings were held to secure an institutional political agreement; group meetings were held with each department or institution's technical professionals, to incorporate the available science and technology information; and political and technical meetings were held to evaluate the portal in its final version.

International examples were used as models, including OpenNASA, LA Referencia, the open-data page of the Massachusetts Institute of Technology and OpenAIRE. These models were chosen because of the large amounts of information handled, their integration and interoperability, the power of their search engines, and their ability to recover and communicate information.

European Union: Horizon 2020 Open Research Data Pilot and Data Management Plan

The initiative was launched by the European Commission's Directorate-General for Research and Development through a broad public consultation on "Science 2.0: Science in Transition" in 2014 (European Commission, 2014). Among the policy interventions discussed was support for data sharing, management, curation and storage. The proposed interventions included building relevant infrastructure, developing data skills, incentivising data sharing and nurturing the development of good practice in handling data (European Commission, 2018a).

Based on the consultation results, it was decided to include an Open Research Data Pilot. Guidance is provided through H2020 Programme guidelines on FAIR data management (European Commission, 2016c) and open access to publications and research data in H 2020 (European Commission, 2017). Article 29.3 of the Grant Agreement for H 2020 formulates the legal requirement. Opt-out from the requirement is possible if the action's main objective would be jeopardised by making those specific parts of the research data openly accessible. In this case, the data-management plan must contain the reasons for not giving access.

Mexico: Open science policy – Open Repositories Programme

The Open Repositories Programme is governed along two main dimensions: legislative and organisational (CONACYT, 2018). The legislative dimension has three levels: federal law, general guidelines and technical guidelines. The same structure applies to the organisational dimension. The Congress of the Union is the only institution authorised to modify the first level; the internal governance structures of CONACYT can modify the second and third-level documents. The first-level legislative document, "Mexican Law for Science and Technology", considers the overall policy goals (Government of Mexico, 2015). The second-level document, "General Guidelines for Open Science", contains guidelines for the policy's design and implementation (Government of Mexico and CONACYT, 2017a). The third-level document, "Specific Guidelines for Open Repositories", considers the more technical elements (Government of Mexico and CONACYT, 2017b). This governance model allows maintaining the main policy goal while updating the technical framework.

The initiative is led by CONACYT, which drafted the Law on Science and Technology. The repositories programme was included in the law as a standalone objective, in the spirit of a "build it and they will come" strategy focusing on funding the repository. Over time, the "capacity-building" component was considered lacking; it is now considered of equal importance as the funding component.

CONACYT established an open science committee (staffed entirely by CONACYT) as the main governance mechanism for the open science policy. The main stakeholders are research institutions and CONACYT public research centres that have developed or hosted an institutional repository. The stakeholders are consulted on second and third-level legislation; their comments are considered for the final decision, which is taken by the committee. When the issue concerns solving a specific and technical issue, an internal analytic group provides the committee with the necessary information to make a decision.

Sweden: Infrastructures for Register-based Research, a government commission to the Swedish Research Council

The Swedish initiative is entirely led by the Swedish Research Council (SRC) and was initially managed according to the SRC project model. The project is governed by an internal steering group, comprising members representing high-level expertise in register-based research, information technology and research infrastructures. The steering group is chaired by the SRC director general. In 2015, the unit for register-based research was established and entrusted with co-ordinating related activities. In 2018, this task was included in the SRC instructions and is now considered permanent (Eriksson and Nilsson, 2018).

At the outset, extensive mapping of the register-based research landscape was conducted in order to understand the bottlenecks and hindrances facing researchers in their projects. It revealed that the main issues lie in the formulation phase, when the researcher has to define the exact data needed for the project – a difficult task without possessing a good understanding of the variables at hand. It was therefore decided to structure the infrastructure on a principle separating the data from the non-sensitive metadata and semantics, which would then facilitate dialogue between the researcher and the register holder (Figure 3.2). The Register Utiliser Tool allows such rich description of the register contents at a variable level, using non-sensitive metadata.

Responsibility for metadata and semantics is assigned to each register owner's organisation, and the infrastructure does not engage in harmonisation activities between the register owners. Rather, the researchers are provided with thorough descriptions of variables, including rich metadata and semantics, to allow them to compare variables and inform a judgement on harmonisation possibilities. The project's work on method and training, together with the infrastructure's application support, aims to promote the use of standardised terminologies, ontologies and classifications, and to make it easy to choose from standards (if appropriate) when working with the metadata and semantics. The choice and decisions are, however, still made by the register owner's organisation.

United States: Principles for Promoting Access to Federal Government-Supported Scientific Data and Research Findings through International Scientific Co-operation

The initiative comes from the White House's Office of Science and Technology Policy (OSTP). Pursuant to the 2013 OSTP directive (Holdren, 2013), each US government funding agency with over USD 100 million (US dollars) in annual conduct of research expenditures developed a plan to increase public access to peer-reviewed scholarly publications and the associated digital scientific data necessary to validate the published research findings. Generally, each agency plan was required to provide strategies for maximising access to, and increasing the findability and reusability of, research results funded by that government agency. In providing such access, agencies were required to address concerns by protecting data: i) relating to personal privacy; ii) subject to intellectual property rights; or iii) that could harm US national, homeland and economic security. In addition, agencies had to balance the benefits associated with long-term preservation and access with the costs and administrative burden of data curation and storage (US Government, 2019).

Bottom-up initiatives driven by institutions

The initiatives in this category are taken by individual institutions grouped under loose consortia. In two of the four initiatives (Netherlands and Colombia), the ministries are involved, albeit not in the driving seat. In the Netherlands, the ministry sets up the National Platform Open Science, but does not participate in the steering committee. In Colombia, the ministry is part of the steering committee, but the co-ordinator is a research institute. In the UK Concordat, government involvement is even more remote, with only verbal support for the initiative and a commitment not to interfere with a consensus achieved by research stakeholders. In Belgium, policymakers are not formally or informally involved in largely operational initiatives.

The following section synthesises issues related to the governance and processes of the four cases in this category.

Netherlands National Plan Open Science

The Open Science Declaration has four key ambitions: i) 100% open access publishing; ii) optimal (re)use of research data; iii) recognise researchers with corresponding evaluation and valuation systems; and iv) encourage and support open science.

It was signed on 9 February 2017 by major research institutions and associations in the Netherlands that have come together in the National Platform Open Science. The platform is supported by the Ministry of Education, Culture and Science, which will organise its set-up, including a secretarial office and website. The signatories include: the Association of Universities in the Netherlands;[8] the Royal Netherlands Academy of Sciences;[9] the Netherlands Organisation for Scientific Research;[10] the Netherlands Association of Universities of Applied Sciences;[11] the PhD Candidates Network of the Netherlands;[12] the National Library of the Netherlands;[13] the Collaborative Organisation for ICT in Dutch Education and Research;[14] the Dutch Federation of University Medical Centres;[15] the Netherlands Organisation for Health Research and Development;[16] and GO (Global Open) Findable, Accessible Interoperable, Reusable (GO FAIR)[17] (Ministry of Education, Culture and Science, 2018).

The governance of the National Plan for Open Science (NPOS) consists of the following:

- **The steering group**, which guards the progress, decides on the direction, ensures commitment and unity, and connects with (inter)national initiatives. It ensures commitment to the NPOS at the highest level. It consists of the executives of selected parties involved in the Open Science Declaration.
- **The National Co-ordinator Open Science**, who serves as a "voice of scientists" and advisor to the steering committee on the key ambitions of the NPOS, represents the National Plan and the National Platform to the outside world, and brings attention to open science at the national and international level. The co-ordinator is preferably a scientist with extensive knowledge in the field of open science.
- **The National Platform Open Science**, whose role is to promote regular deliberation between the parties involved in the Open Science Declaration, to ensure that the ambitions of the NPOS are realised, and that coherence and co-operation in the field of open science is stimulated. It consists of open-science experts from the parties involved in the Open Science Declaration.
- **Theme groups**, which lead in-depth thematic discussions on the four key ambitions of the NPOS to realise and implement the ambitions, and inform the Platform on progress. They consist of open-science experts from the parties involved in the Open Science Declaration.
- **The Secretariat**, which supports the Platform chair (Ministry of Education, Culture and Science) both in terms of logistics and content, serves as the contact point for Dutch and international parties, and assists in carrying out platform activities.

Following the adoption of the Declaration in 2017, the theme groups, in co-operation and consultation with the steering committee, national co-ordinator and platform, have been working on implementing the NPOS. They have produced a Roadmap to open access, as well as recommendations for recognising and rewarding researchers for furthering open science.

UK Concordat on Open Research Data

The UK Concordat on Open Research Data (UK Research and Innovation, 2016) was initiated by the UK Open Research Data Forum, a body established in January 2014 by the Royal Society. The Concordat's membership comprises representatives from academies, universities, funders, publishers, private-sector research (pharmaceuticals) and key related bodies supporting research, such as Jisc, the Open Data Institute and the British Library. The Open Research Data Forum proposed the development of the Concordat to help accelerate culture change. A draft concordat was released for consultation in 2015. About 80 responses were received from a variety of organisations; these helped inform the final version published in 2016, featuring signatories from research funders and universities, specifically Universities UK, Research Councils UK (RCUK), the Higher Education Funding Council for England (HEFCE) and the Wellcome Trust (Bruce, 2018).

The then Minister of State for Universities, Science, Research and Innovation wrote a supportive foreword clarifying the status of the Concordat: "This is not a government-owned document, nor should it be. The research community has worked hard to arrive at the consensus delivered in this report and I would like to thank the members of the UK Open Research Data Forum for their valuable contributions."

The Concordat has had further signatories. In discussion with other stakeholders and signatories, its ongoing governance and implementation fall under UK Research and Innovation.[18]

The Concordat built on the legacy of the OECD *Recommendation of the Council concerning Access to Research Data from Public Funding* (OECD, 2006), the Research Councils' Common principles on Data (Research Council UK, 2012) and the report by The Royal Society (2012) on "Science as an Open Enterprise", as well as of European policies such as the Commission Recommendation 2012/417/EU on access to and preservation of scientific information (European Union, 2012) and the "G8 science ministers statement: London" (G8, 2013).

South Africa: DataFirst repository

DataFirst's repository is located within the University of Cape Town's governance structure. Funded partly by the university and partly by grants, the independent unit is based at the Faculty of Commerce. DataFirst's director, who is also a full-time professor at the School of Economics, has overall responsibility for the unit and DataFirst research initiatives. DataFirst's manager oversees the operations of the repository and other data services, and supervises data-service staff (Woolfrey, 2018).

A governing board provides oversight. It meets annually to review DataFirst's annual report, provide input on operations and discuss relevant scientific developments. The board comprises representatives from South Africa's Department of Science and Technology, National Statistics Agency and other government departments, and research-intensive universities, as well as from the African Economic Research Consortium (a policy think-tank). International board members include the directors of two well-established data archives: the UK Data Archive at the University of Essex and the Inter-university Consortium for Political and Social Research at the University of Michigan. DataFirst's strategic goal is to promote the efficient and equitable sharing of South Africa's public-sector data for research purposes. Its governance model allows clients in government and academia to provide feedback on adherence to this goal. DataFirst's board also advises on developments in data science and policy.

Belgium: Data Management Plan Belgium Consortium

Data Management Plan Belgium Consortium is led by a consortium of all Belgian universities, except KU Leuven, which has its own DMPonline programme. DMPonline.be is a joint service for which partners are jointly responsible (i.e. it is not a software product or service for sale) on a best-effort basis, rather than with a service-level guarantee (Laureys, 2018).

The general assembly, in which each partner is represented, manages the project and consortium; Ghent University acts as the co-ordinator. The assembly can take valid decisions with a two-thirds quorum and a two-thirds majority of votes cast, although some decisions require unanimous votes or a positive vote from the co-ordinator. One policymaker from the Federal Science Policy Office is an observing member, and close ties are maintained with the Flemish Interuniversity Council (VLIR) and French-speaking Rectors' Council.

During the pilot phase in 2015-16, Ghent University Library installed a local version of the open-source DMPonline software (version 4.1), an application for data-management planning originally developed by the United Kingdom's Digital Curation Centre. Possibilities for collaboration were discussed in a number of settings, e.g. at the VLIR working group on research-data management, at open-access networking events and at several meetings of university libraries.

In 2017, the consortium was formalised through a draft consortium agreement that also outlined how new partners can enter the consortium. The consortium also agreed on a code of conduct for people involved in administering the software. At the same time, the software was moved to Belnet servers and enriched with a number of features. In 2018, more institutions joined the consortium, including four French-language universities and a public research organisation. A final consortium contract is yet to be signed.

Colombia: Colombian Biodiversity Information System – SiB Colombia

The governance structure of SiB Colombia is made up of a partners' network. Partners can be individuals, organisations or networks, who assume one of three possible roles: i) data publisher, contributing data to the system; ii) advisor, contributing expertise; or iii) promoter, helping to diffuse information on SiB Colombia, build capacity, and/or secure economic or other resources for the operation of SiB Colombia (Escobar, Hernández and Agudelo, 2018).

The network is organised around the following bodies:

- **Steering committee (CD-SiB)**: the steering committee represents partners' interests, evaluates operating guidelines, and orients strategies and resources. Its members include the Ministry of Environment and Sustainable Development, the directors of five leading research institutes in various fields related to biodiversity, the rector of the Universidad Nacional de Colombia and the director of the National Natural Parks.
- **Technical committee (CT-SiB)**: the technical committee defines the conceptual and operational guidelines that guide the scope and activities of SiB Colombia, consolidates tools, products and information services to improve access to data and information on biodiversity in Colombia; validates data; and performs other technical tasks. Its members are delegates from the institutions represented in the steering committee, complemented by invited experts as needed.
- **Interest and work groups**: these groups work on subjects related to biodiversity, in response to the specific needs of SiB Colombia.
- **Co-ordinating team (EC-SiB)**: the co-ordinating team is a group of professionals who ensure the coherence and viability of both SiB Colombia's infrastructure assembly, start-up and maintenance processes, as well as its content, governance and communication.

The Alexander von Humboldt Biological Resources Research Institute is the entity responsible for the design, implementation and co-ordination of the country's SiB. In accordance with Law 99 of 1993, and regulatory decrees 1600 and 1603 of 1994, the Institute houses the EC-SiB.

Queries to stakeholders are conveyed through the EC-SiB and its co-operation line, which leads the design and implementation of the strategies and processes promoting the expansion and consolidation of the SiB Colombia network. The EC-SiB is responsible for facilitating the implementation of the governance guidelines; maintaining the spaces and mechanisms enabling the participation of the people, organisations and networks comprising SiB Colombia; and co-ordinating and facilitating the collaborative construction of projects.

International influences on national policy making

Several international frameworks are shaping national policy agendas in the domain of enhanced access to and sharing of data (see Chapter 2). Many of the case studies mention these frameworks' influence on their own policy making. The interconnectedness of STI systems makes it almost inevitable to consider such references. The case studies mention OECD, Group of Eight (G8), UNESCO and European Commission initiatives; the work of the Committee on Data (CODATA) of the International Council for Science and the Research Data Alliance; and reference national policy making in the United States, the United Kingdom and Australia; the OECD *Recommendation of the Council concerning Access to Research Data from Public Funding* (OECD, 2006), as well as OECD work on open science (OECD, 2015); the UNESCO Policy Guidelines for the Development and Promotion of Open Access (Swan, 2012); and the G8 Science Ministers Statement (G8, 2013).

The European initiatives that are most often mentioned include the Commission Recommendation 2012/417/EU on access to and preservation of Scientific Information (European Commission, 2012), the GO FAIR initiative and the European Open Science Cloud (European Commission, 2018b). The case studies point to free movement of information as the fifth EU freedom (after the free movement of people, goods, services and capital). They also mention the Budapest Open Access initiative when policies have a combined scope, covering open access to both scientific publications and data. Countries where access to data for STI is linked to open-government frameworks quote the Open Government Partnership[19] as relevant to access to research data. South Africa's DataFirst initiative reports the most comprehensive benchmarking with international frameworks (Table 3.2).

International interoperability is key. For example, Finland considered internationally interoperable metadata as an important factor, not only for operating the services, but also obtaining international visibility and achieving global impact (Haapamaki, 2018). The Slovenian open-access strategy states that "the national open-access infrastructure for peer-reviewed publications and research data has to be interoperable with relevant European and international infrastructures. Slovenian publications repositories, research-data repositories and archives as well as software for scientific journal publishing have to be compatible with OpenAIRE guidelines, which will enable the European and national research funders to monitor the compliance with open-access mandates for scientific information." (Tramte, 2018) Although it is originally European, Argentina and Mexico have also adopted OpenAIRE for their repositories (Luchilo, D'Onofrio and Tignino, 2018).

Several countries are also actively involved in shaping such initiatives: Germany, the Netherlands and France are founding members of GO FAIR. All EU member states are gearing up for the European Open Science Cloud, and also contribute to EU recommendations and directives.

France is pioneering an approach to extend data sharing to private-sector data of public interest. This is already enacted in French legislation, and President Macron wants to co-ordinate such an approach at the European level (Présidence de la République, 2018). France also wishes to propose new forms of data sharing for AI through public-private partnerships (DGE, 2018) and to create an intergovernmental expert group to further develop these ideas (French Ministry of Higher Education, Research and Innovation, 2018).

Further regional approaches include Mexico's role in promoting open science as regional leader of the Ibero-American Summit, and helping Panama and Paraguay design their own national open-science policies (CONACYT, 2018). Sweden has also led an effort which led to the adoption of, or mapping to, the General Statistical Information Model standard in the Nordic statistical institutes, thereby increasing interoperability on a Nordic scale (Eriksson and Nilsson, 2018).

International co-operation itself is an ingredient of open science as a global phenomenon. It materialises, for example, through the Global Biodiversity Information Facility (GBIF), an international network and research infrastructure funded by the world's governments and aiming to provide open access to data about all types of life on Earth to anyone, anywhere. SiB Colombia is the "node" responsible for co-ordinating GBIF-related activities in Colombia since 2003 (Escobar, Hernández and Agudelo, 2018).

Table 3.2. Commonalities among data-sharing principles

Principles used as references by the DataFirst repository at the University of Cape Town, South Africa

Date	Name	Data	Accessibility				Interoperable	Machine-readable	Timely	Secure	Interpretable	Permanent
			Discoverable	Non-discriminatory	Complete	Primary						
			Data should be online and easy to find	Access should be on equal terms	Data should be made available in their entirety	Primary, not aggregate data	Standardised open formats	Structured to allow automated processing	Shared in a timely manner	Protection of privacy of data subjects	Good data documentation	Preserved and shared in the long term
1996	Bermuda Principles	Genomics data		o	o				o			
1997	Data quality principles	Government data			o	o			o	o	o	o
2007	OECD Principles and Guidelines for Access to Research Data from Public Funding	Data from publicly funded research	o	o	o	o	o		o	o	o	
2007	Sebastopol Principles - eight fundamental principles for open government data	Government data	o	o	o	o	o	o	o		o	
2010	10 Open Government	Government data	o	o	o	o	o	o	o			o
2014	FAIR Data Principles (findable, accessible, interoperable, re-usable)	Research data	o	o			o	o			o	o
2015	ICSU-WDS Data Sharing Principles	Research data							o	o		

Source: Woolfrey (2018), "The DataFirst Research Data Repository", https://community.oecd.org/servlet/JiveServlet/downloadBody/149008-102-2-264196/20190228-oecd-case-study-sa-df-v2.pdf.

Monitoring and evaluation of policies

Although monitoring of the initiatives could be improved, some positive impacts can be identified. Overall, few of the case studies have developed specific monitoring schemes, although several feature such objectives in their roadmap. Even fewer have specific impact assessment frameworks, whether ex ante or ex post.

Finland, which has developed a comprehensive evaluation system, is a positive example of monitoring and evaluation. The strategy group outlined a set of key indicators for monitoring progress on the ATT initiative. An independent review was conducted in 2016, analysing a range of developments, from the level of international policies to the "grassroots". Individual and group interviews, as well as a web survey, were used to investigate issues such as cultural change towards openness, perceived benefits and drivers of the transition towards openness. The report further investigates how researchers harness the benefits of open science and research, the perceived societal benefits, challenges and the way forward (Tuomi, 2016). In addition, the report evaluates openness in the activities of research institutions and research-funding organisations, highlighting best practices and areas of development, while initiating discussions on open science and research at the international level (Haapamaki, 2018).

The European Commission also has an advanced monitoring system, at the level of compliance with both its Recommendation on access to and preservation of scientific information, and its Open Research Data Pilot and Data Management Plan (Box 3.3) (European Commission, 2018a).

In Germany, the agreement between the federal government and the *Länder* includes the evaluation of the effectiveness of funding as well as the structural effects of the National Research Data Infrastructure. The intended evaluation is twofold: it addresses the effectiveness of both the National Research Data Infrastructure itself and the funded consortia. The consortia will be regularly evaluated by the German Research Foundation, while the National Research Data Infrastructure will be structurally evaluated by the Scientific Council after seven years (Philipps and Bodmann, 2018).

Box 3.3. Monitoring implemented by the European Commission

Since the publication of the Recommendation on access to and preservation of scientific information in 2012, the European Commission has been monitoring the impact of the Recommendation, and progress towards the objectives, through the reporting of the Member States and three Associated Countries, thanks to the national points of reference.

Implementation of FAIR principles is uneven across the European Union. As of 2016, 8 countries[1] (out of 31 surveyed) included open access to research data within their research-policy documents (and only 1 had actually implemented it). Approximately one-half of the EU member states impose data-management plans and FAIR principles; other countries sometimes have open-access schemes at the institutional level. Funding for research-data management is available in 12 countries (out of 31), but open research data are eligible under most national funding schemes. In many countries, research-data management is only applied to EU projects.

FAIR research-data management practices (focusing on various different research artefacts produced as part of scientific activities, e.g. datasets, software tools, workflows, notebooks, ontologies and articles) need to be developed and implemented at the national level, with more room for work on TDM, its effects on research and positive conclusions regarding developments in e-infrastructure policy.

Figure 3.2. National policies or overall strategies to encourage or mandate dissemination of and open access to research data are defined at the national level

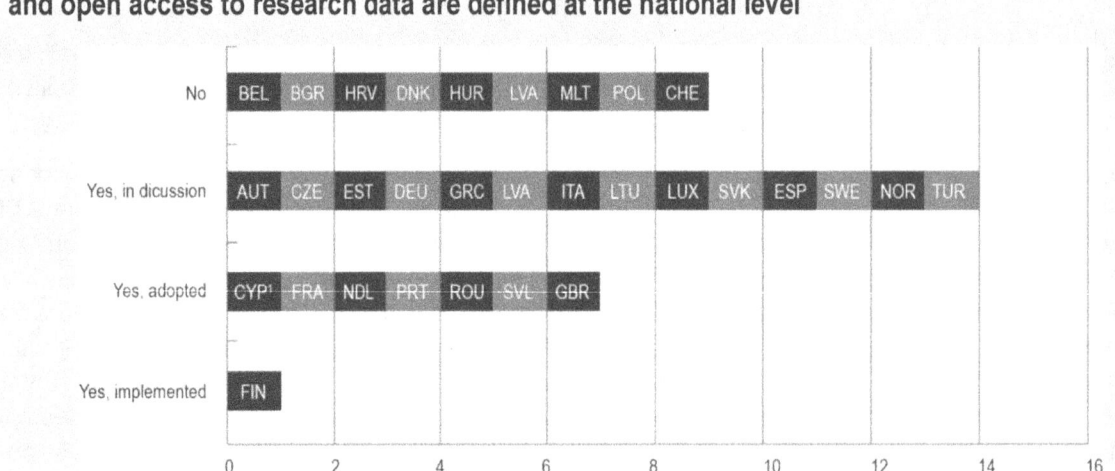

1. Note by Turkey:
The information in this document with reference to "Cyprus" relates to the southern part of the Island. There is no single authority representing both Turkish and Greek Cypriot people on the Island. Turkey recognises the Turkish Republic of Northern Cyprus (TRNC). Until a lasting and equitable solution is found within the context of the United Nations, Turkey shall preserve its position concerning the "Cyprus issue".
Note by all the European Union Member States of the OECD and the European Union:
The Republic of Cyprus is recognised by all members of the United Nations with the exception of Turkey. The information in this document relates to the area under the effective control of the Government of the Republic of Cyprus.
Source: European Commission (2018a), "Case study of policy initiative for open access to research data: Horizon 2020 open research data (ORD) pilot and data management plan", https://community.oecd.org/servlet/JiveServlet/downloadBody/141323-102-1-248452/European_Commission.pdf.

Study for the Open Research Data Pilot in 2017

In 2014-16, 68% of the funded projects in the core areas participated in the Open Research Data Pilot. Correspondingly, the average opt-out rate in signed grant agreements was 32%. Outside the core areas, 9% of projects availed themselves of the voluntary opt-in possibility. It is notable that as the sample size increases from the first datasets (2014-15) to the second dataset (2015-16), the opt-outs decrease, and voluntary opt-ins in the non-core areas increase. This points to the policy's overall success, both in terms of percentage and absolute number of projects participating.

In 2017, when the Open Research Data Pilot was extended to all areas of H 2020, initial data suggest a slightly lower participation rate (62.09%) and a slightly higher opt-out rate (37.91%). However, these are data is based on a small sample and needs to be corroborated by further evidence covering 2018.

The reasons for opting out remain consistent in the two datasets. Intellectual property (IP) comes first (46%), followed by no data generated (projects concern co-ordination and support actions) (17.9%) and privacy/personal-data protection (17.6%); a further 8.1% of applicants state they opted out because opening up research data would hinder their achieving the project's main objective, and 5.3% opted out owing to confidentiality in a national-security context.

1. The latest report, published in 2018, gives the status as of 2016. Since then some countries, such as Norway and Denmark, have adopted strategies.
Source: European Commission (2018a), "Case study of policy initiative for open access to research data: Horizon 2020 open research data (ORD) pilot and data management plan", https://community.oecd.org/servlet/JiveServlet/downloadBody/141323-102-1-248452/European_Commission.pdf.

The Norwegian, Slovenian and Korean strategies do not yet have specific monitoring frameworks. However, in Korea, the investment plans for the strategy will undergo a review that systematically assesses the scientific, economic and policy impacts of the investment plans, and decides whether to go ahead with the investment or not (Shin, 2018). Norway will develop relevant statistics on the generation and sharing of research data (Norwegian Ministry of Education and Research, 2018). In Slovenia, the Ministry of Education, Science and Sport has not yet established monitoring and evaluation processes, which will be created during the initial phases of the implementation (Tramte, 2018).

France foresees the development of an open-science barometer in the future as part of the recently launched National Plan for Open Science (French Ministry of Higher Education, Research and Innovation, 2018). In the Netherlands, monitoring of open access also remains to be developed as one of the five pillars in the Roadmap to Open Access 2018-2020 (VSNU, 2018; Ministry of Education, Culture and Science, 2018).

Principle 10 of the UK Concordat's states that "[r]egular reviews of progress towards open research data should be undertaken." It further calls for reviews to be regular, "not over-burdensome but rather flexible and recognise that developments will take time. Their essence should be one of identifying and sharing best practice. This would be best achieved through engagement with community activities, such as the UK Open Data Forum, that bring together the full range of stakeholders." (Bruce, 2018)

In addition to general open-government data indicators, Canada has developed a dedicated framework for evaluating open-science implementation which uses the statistics of open-access datasets, as well as altmetrics attention scores and qualitative descriptions of open science-related software (Government of Canada, 2018; Treasury Board of Canada Secretariat, Open Government Team, 2018).

The Colombian Biodiversity Portal collects basic statistics such as: i) number of entities and people publishing data per year; ii) number of visits and download of data from SiB Colombia web portals per month; iii) number of published data per month; and iv) type of data (taxonomic and geographic) (Escobar, Hernández and Agudelo, 2018). The Argentine Science and Technology Information Portal also does not have a specific framework and relies on Google Analytics to monitor portal activity (Luchilo, D'Onofrio and Tignino, 2018). South Africa's DataFirst combines usage metrics (e.g. client registrations and data downloads) and publication citation counts, and also uses Google Analytics (Woolfrey, 2018). In Mexico and Sweden, monitoring is based on the funding agency's usual procedure (CONACYT, 2018; Eriksson and Nilsson, 2018).

Achievements and challenges

Achievements

The initiatives presented in the case studies have served the purpose of advancing the agenda of enhanced access to data for STI – and more generally the open-science agenda – through increased awareness and consensus-building among stakeholders. Depending on the degree of maturity, their positive is more qualitative or quantitative.

Among the high-level initiatives presented, Finland's ATT initiative is the most mature. It has demonstrated its impact as an accelerator of open science both in Finland and in the international context. It has been able to address a number of issues, such as digitalised services for the research field, creating reference architecture for open science, providing practical guidelines and support for researchers, and creating models and tools for open access and long-term preservation of metadata. In addition, the ATT initiative has provided comparative information allowing funding and research organisations to determine their position regarding open science (Haapamaki, 2018).

The "future university" was an overwhelmingly popular topic in the ATT impact assessment, including how open science changes the way research is conducted and how the whole sector needs to embrace openness. The assessment also identified the need to look at open science from the perspective of wider audiences: for instance, how are students involved and what kinds of open-science competences will they need in the future? Some respondents noted that open science should be taught not only in higher education, but also at schools and kindergartens. Respondents also highlighted the importance of the operator's role, on the premise that leaving the transformation to individual organisations would risk slowing down the pace of change. As so many major questions remain unanswered, the ATT initiative should be continued in some form.

Other high-level initiatives – such as Norwegian, Korean and Slovenian strategies, the Netherlands' Open Science Policy and the UK Concordat – have yet to be implemented, and quantitative impact is therefore expected in the future. To date, they have had more qualitative achievements, such as raising community awareness and initiating dialogue about data sharing among key stakeholders (academia, policy makers, data repositories and the private sector). The UK Concordat concludes it has helped ensure that university leaders are aware of the needs for Open Research Data, its benefits and the key challenge areas. It has ensured initial buy-in and commitment, and resulted in agreed principles for and high-level endorsement of Open Research Data (Bruce, 2018).

In Korea, the debate on open access was linked to unleashing the power of big data and data-driven innovation during the Fourth Industrial Revolution (Shin, 2018). In Argentina, consensus-building around the portal has helped partly overcome the cultural barrier against data sharing (Luchilo, D'Onofrio and Tignino, 2018).

As pointed out by Slovenia (but valid in most cases), provisions for research-data management and openness are not designed as standalone solutions, but are fully aligned with international recommendations. Also, the national approach to designing provisions is more efficient than the introduction of separate provisions by individual research-funding and research-performing organisations (Tramte, 2018).

Adoption of a common standard and a technical framework has also been reported by both Mexico (OpenAIRE) (CONACYT, 2018) and Sweden (Generic Statistical Information Model [GSIM]). Sweden reports the GSIM acts as an effective common language of metadata and provides a framework for identifying what parts of the metadata and semantic descriptions need to be further curated, and which parts can be harvested and used "as is" (Eriksson and Nilsson, 2018). However, Colombia points out the GSIM common standard is a necessary support framework, but was not the strategy underpinning the construction of an initiative such as SiB Colombia (Escobar, Hernández and Agudelo, 2018).

As explicitly reported by Colombia (but implicitly stated in many other case studies), a set of principles and values is the basis for the trust necessary to build the initiative. The construction of an open-access ecosystem is a collective exercise: recognition and visibility are the first steps required to generate confidence. The discourse and the language must be coherent with the principles and values, and in turn with the actions. Additionally, it is important to build on the available infrastructure, rather than starting from a blank sheet.

The European Open Research Data Pilot shows more quantitative impact, with a relatively high opt-in rate (68%) – demonstrating that researchers are willing to share their data in most cases. The researchers who opted out (32%) quoted IP of data as the major reason. This finding would need additional analysis, to understand whether it relates to specific projects combining proprietary background knowledge with research financed by H 2020 (European Commission, 2018a).

Among the case studies presented, SiB Colombia, initiated 24 years ago, has the longest track record. This initiative shows quantified achievements, including 2.45 million records referring to 62 829 species, 101 publishing institutions, 5 340 species profiles and 130 species checklists. In addition, SiB Colombia reports the following achievements: i) implementation of an open-source platform[20] ii) creation of an open-access guide and data-use policy;[21] iii) development of a robust and interoperable data-publishing model;[22] and iv) establishment of a solid and operational governance model[23] (Escobar, Hernández and Agudelo, 2018).

Other initiatives are still in the build-up stage and already have an impact, which can be further developed. For example, the Argentine portal launched in 2017 now contains 18 000 projects from the main funding sources, but still needs to include projects from other scientific organisations and universities (Luchilo,

D'Onofrio and Tignino, 2018). Mexico reports ten ongoing Open Research Data repositories, with 27 000 research datasets (CONACYT, 2018). Canada's Open Government Portal allows users to search through over 81 000 open data and information from departments and agencies across the federal government (Treasury Board of Canada Secretariat, Open Government Team, 2018). The Swedish infrastructure is operational, with a limited number of sensitive register datasets. However, the SRC, Statistics Sweden, and the National Board of Health and Welfare are currently discussing the need to devise technical solutions for data disclosure that can also link register data with big data, as well as the need to expand the dialogue to other actors in the e-infrastructure ecosystem (Eriksson and Nilsson, 2018).

Future challenges

Cultural barriers remain a cross-cutting challenge. They stem from researchers' reluctance to share data, owing to concerns over the associated risks and burdens. The attitude of Korean researchers is illustrated in Figure 3.3, but most case studies mention similar attitudes. Slovenia reports that recognition of efforts to ensure Open Research Data in research-evaluation processes would encourage researchers to embrace openness more quickly (Tramte, 2018). Mexico reports that institutions participate in the Open Repositories Programme, but that the academic community does not (CONACYT, 2018). The UK Concordat points out that cultural change will take time (Bruce, 2018). As Norway correctly notes, it is important not to have unrealistic expectations. While there exist many legitimate reasons for demanding confidentiality of research data, a superficial disregard of open-access obligations should not be accepted when such issues are raised. Further, a real judgement should be made of how data can be made "as open as possible" within those constraints (Norwegian Ministry of Education and Research, 2018).

Figure 3.3. Korean researcher perceptions of open (research) data

Percentage of responding researchers who had disclosed their data and underlying reasons

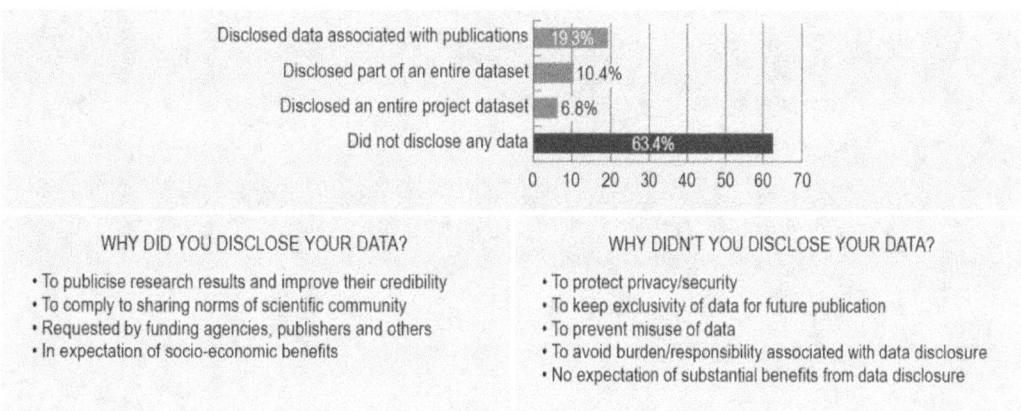

Source: An et al. (2017), "Survey of Korean researchers on open science", recited from Shin (2018), "Korean case report on enhanced access to public data for science, technology & innovation", https://community.oecd.org/servlet/JiveServlet/downloadBody/141310-102-4-263210/korean%20case%20report.pdf.

Another cross-cutting issue is the **lack of appropriate skills** in the build-up to projects. In the example of the Argentine portal, the technical team had to self-train, through virtual courses and a trial-and-error approach (Luchilo, D'Onofrio and Tignino, 2018). Finland anticipates insufficient or poorly organised resources for development and training, mostly owing to a lack of long-term funding (Haapamaki, 2018). Slovenia reports a lack of data experts in the country, slowing down the implementation of its open-access strategy (Tramte, 2018). Sweden underlines the importance of involving all relevant competences, spanning metadata and semantics, and understanding the research domain to adapt to researchers' needs. In addition, researchers' skills in research-data management and openness need to be further developed (Eriksson and Nilsson, 2018).

It is important to understand the domain and **researchers' underlying needs**. For example, modelling concept systems that describe the semantics of the concepts used in the register context is fundamental to understanding the content. If done properly, such modelling allows researchers to use data without delving into the technical details or understanding complex variable names.

Research-data management **plans** are perceived as an important instrument in emphasising data considerations early in the project cycle, but requirements need to be balanced against the administrative burden on researchers and the need for flexibility in different types of research. Research-data management is not always budgeted for and thus comprises additional burden to researchers, which is not necessarily compensated. The next European Framework Programme will require research-data management plans, even for projects that opt out from opening their data (European Commission, 2018a).

Moreover, Norway points out the need to recognise the **significant costs of open access**, and to reduce those costs and find business models that can sustainably uphold open-access efforts (Norwegian Ministry of Education and Research, 2018). Mexico reports that its national open science policy lacks its own resources, and that its funding therefore depends on another programme (CONACYT, 2018).

The rich variety of policy initiatives reviewed in the various case studies illustrates the different approaches to enhancing data access. The issue is being given strong policy attention, and in many cases there is a strong top-down effort to create a comprehensive strategy, striking the appropriate balance between reaping the benefits of enhanced data sharing and managing the associated risks. The next chapter will address the remaining policy gaps more specifically.

References

An et al. (2017), "Survey of Korean researchers on open science", paper presented at the Korean Society for Innovation Management and Economics (KOSIME), 23-24 June 2017, Jeju, Korea (unpublished).

Bruce, R. (2018), "UK case study: The concordat on open research data", case study for the OECD project on enhanced access to data, https://community.oecd.org/servlet/JiveServlet/downloadBody/149486-102-1-264203/OECD-open-access-to-data-UK-Case-Study-v01_02%20(002).pdf.

BSN (26 January 2018), "Evolution de la BSN vers le Comité pour la Science Ouverte (CoSO)", CoSO news blog, https://www.ouvrirlascience.fr/evolution-de-la-bsn-vers-le-comite-pour-la-science-ouverte-coso/ (accessed on 26 February 2020).

CONACYT (2018), "Mexican open science policy – Case study: Open institutional repositories program", case study for the OECD project on enhanced access to data, https://community.oecd.org/servlet/JiveServlet/downloadBody/149104-102-1-263395/Mexico.pdf.

DAMVAD Analytics A/S (2014), *Sharing and Archiving of Publicly Funded Research Data: Report to the Research Council of Norway*, DAMVAD, Oslo, Copenhagen, https://www.usit.uio.no/om/organisasjon/itf/saker/forskningsdata/bakgrunn/sharing-and-archiving-research-data.pdf (accessed on 26 February 2020).

DGE (2018), "Mutualisation de données pour l'intelligence artificielle", webpage, Direction générale des entreprises (DGE), https://www.entreprises.gouv.fr/numerique/mutualisation-de-donnees-pour-intelligence-artificielle (accessed on 4 October 2019).

EC/OECD (2018), *STIP Compass*, https://stip.oecd.org/stip.html (accessed on 30 August 2018).

Eriksson, M. and M. Nilsson (2018), "OECD case study report RUT", case study for the OECD project on enhanced access to data, https://community.oecd.org/servlet/JiveServlet/downloadBody/141329-102-1-248465/Sweden.pdf.

Escobar, D., A. Hernández and M. Agudelo (2018), "SiB Colombia – case study: Enhanced access to public data for science, technology and innovation", case study for the OECD project on enhanced access to data, https://community.oecd.org/servlet/JiveServlet/downloadBody/149102-102-1-263390/SIB%20Colombia%20-%20Case%20Study%20open%20access-AP.pdf.

Etalab (2018), "Plan d'action gouvernement ouvert 2018-2020", webpage, Direction interministérielle du numérique (DINUM), France, https://gouvernement-ouvert.etalab.gouv.fr/pgo-concertation/topic/5a1bfc1b498edd6b29cb10d4 (accessed on 3 October 2019).

European Commission (2018a), "Case study of policy initiative for open access to research data: Horizon 2020 open research data (ORD) pilot and data management plan", case study for the OECD project on enhanced access to data, https://community.oecd.org/servlet/JiveServlet/downloadBody/141323-102-1-248452/European_Commission.pdf.

European Commission (2018b), "EOSC strategic implementation roadmap", presentation, European Commission, Directorate General for Research and Innovation, May, https://ec.europa.eu/research/openscience/pdf/eosc_strategic_implementation_roadmap_short.pdf#view=fit&pagemode=none (accessed on 3 July 2019).

European Commission (2017), *Guidelines to the Rules on Open Access to Scientific Publications and Open Access to Research Data in Horizon 2020*, European Commission, Directorate-General for Research and Innovation, https://ec.europa.eu/research/participants/data/ref/h2020/grants_manual/hi/oa_pilot/h2020-hi-oa-pilot-guide_en.pdf (accessed on 28 February 2020).

European Commission (2016a), *Open Innovation, Open Science, Open to the World – A Vision for Europe*, European Commission, Directorate-General for Research and Innovation, https://ec.europa.eu/digital-single-market/en/news/open-innovation-open-science-open-world-vision-europe (accessed on 26 February 2020).

European Commission (2016b), *"European Cloud Initiative – Building a Competitive Data and Knowledge Economy in Europe"*, communication, European Commission, https://ec.europa.eu/digital-single-market/en/news/communication-european-cloud-initiative-building-competitive-data-and-knowledge-economy-europe (accessed on 26 February 2019).

European Commission (2016c), *H2020 Programme – Guidelines on FAIR Data Management in Horizon 2020*, European Commission, Directorate-General for Research and Innovation, http://ec.europa.eu/research/participants/data/ref/h2020/grants_manual/hi/oa_pilot/h2020-hi-oa-data-mgt_en.pdf (accessed on 28 February 2020).

European Commission (2014), "Validation of the results of the public consultation on Science 2.0: Science in transition", policy brief, European Commission, Directorate-General for Research and Innovation, http://ec.europa.eu/research/consultations/science-2.0/science_2_0_final_report.pdf (accessed on 26 February 2020).

European Commission (2012), *Commission Recommendation 2012/417/EU on access to and preservation of scientific information,* Official Journal of the European Union, European Commission, July, https://op.europa.eu/en/publication-detail/-/publication/48558fc9-d4c8-11e1-905c-01aa75ed71a1/language-en.

European Union (2012), Legislation L 194, Official journal of the European Union, Vol. 55, July, Luxembourg, http://dx.doi.org/10.3000/19770677.L_2012.194.eng.

FECYT (2019), "Spanish law of science, technology and innovation", case study for the OECD project on enhanced access to data, https://community.oecd.org/servlet/JiveServlet/downloadBody/148992-102-1-263212/Spanish%20Law%20of%20Science%2C%20Technology%20and%20Innovation%20v2.pdf.

French Government (2018), "Discours du Président de la République sur l'intelligence artificielle" [speech on artificial intelligence by the President of the Republic], Paris, www.elysee.fr/declarations/article/transcription-du-discours-du-president-de-la-republique-emmanuel-macron-sur-l-intelligence-artificielle (accessed on 4 October 2019).

French Ministry of the Economy, Industry and the Digital Sector (2015), "Projet de loi pour une République numérique" [draft law for a digital Republic], https://www.republique-numerique.fr/media/default/0001/02/ce21a30ba6d31b99c71311438a172e3c547c9dca.pdf (accessed on 2 October 2019).

French Ministry of Higher Education, Research and Innovation (n.d.), *Explore the World of French Research and Innovation with ScanR*, search engine, https://scanr.enseignementsup-recherche.gouv.fr/ (accessed on 27 February 2020).

French Ministry of Higher Education, Research and Innovation (2018), "Ouverture des données publiques et de recherche", case study for the OECD project on enhanced access to data, https://community.oecd.org/servlet/JiveServlet/downloadBody/141327-102-2-263394/French%20Case%20Study_v13.pdf.

G8 (2013), "G8 science ministers statement: London, 12 June 2013" –Department for Business, Innovation and Skills, Prime Minister's Office, and the Rt Hon David Willets, United Kingdom, https://www.gov.uk/government/publications/g8-science-ministers-statement-london-12-june-2013 (accessed on 3 October 2019).

German Federal Government and German Bundesländer (2018), *Bund-Länder-Vereinbarung zu Aufbau und Förderung einer Nationalen Forschungsdateninfrastruktur (NFDI) vom 26. November 2018*, https://www.gwk-bonn.de/fileadmin/Redaktion/Dokumente/Papers/NFDI (accessed on 26 February 2020).

Government of Canada (2018), "Monitoring open science Implementation in federal science-based departments and agencies: Metrics and indicators", report of the science-based departments and agencies (SBDAs) Open Science Metrics Working Group, https://ecccdocs.techno-science.ca/documents/ECCC_STB_STSD_OpenScienceMetricsReportADMOvf-accessible.pdf (accessed on 2 March 2020).

Government of Mexico (2015), *Mexican Law for Science and Technology*, http://www.diputados.gob.mx/LeyesBiblio/pdf/242_081215.pdf (accessed on 2 March 2020).

Government of Mexico and CONACYT (2017a), *General Guidelines for Open Science*, www.siicyt.gob.mx/index.php/normatividad/conacyt-normatividad/programas-vigentes-normatividad/lineamientos/lineamientos-generales-de-ciencia-abierta/4707-lineamientos-generales-de-ciencia-abierta/file (accessed on 2 March 2020).

Government of Mexico and CONACYT (2017b), *Specific Guidelines for Open Repositories*, www.siicyt.gob.mx/index.php/normatividad/conacyt-normatividad/programas-vigentes-normatividad/lineamientos/lineamientos-especificos-para-repositorios/4704-lineamientos-especificos-para-repositorios/file (accessed on 2 March 2020).

Government of the Netherlands (2016), "Amsterdam call for action on open science", document based on input from the April 2016 Amsterdam Conference "Open Science – From Vision to Action", https://www.government.nl/binaries/government/documents/reports/2016/04/04/amsterdam-call-for-action-on-open-science/amsterdam-call-for-action-on-open-science.pdf (accessed on 12 September 2019).

Haapamaki, J. (2018), "Review of the Finnish ATT initiative", case study for OECD project on enhanced access to data, https://community.oecd.org/servlet/JiveServlet/downloadBody/149009-102-1-263236/Report%20of%20the%20Finnish%20ATT%20Initiative_%20versio%201_0_en.pdf.

Holdren, J. (2013), "Increasing access to the results of federally funded scientific research", Memorandum for the Heads of Executive Departments and Agencies, 22 February, Executive Office of the President, Office of Science and Technology Policy, Washington, DC, https://obamawhitehouse.archives.gov/sites/default/files/microsites/ostp/ostp_public_access_memo_2013.pdf (accessed on 14 February 2020).

Laureys, E. (2018), "Belgian case study on open access to data: DMP Belgium consortium", case study for the OECD project on enhanced access to data, https://community.oecd.org/servlet/JiveServlet/downloadBody/141322-102-1-248450/Belgium.pdf.

Luchilo, L., M. D'Onofrio and M. Tignino (2018), "Case study: The Argentine science and technology information portal", case study for the OECD project on enhanced access to data, https://community.oecd.org/servlet/JiveServlet/downloadBody/141311-102-1-248448/Argentina.pdf.

Morais, R. and L. Borrell-Damian (2018), "Open access – 2016-2017 EUA Survey results", report, European University Association, www.eua.be/Libraries/publications-homepage-list/open-access-2016-2017-eua-survey-results (accessed on 19 June 2019).

Netherlands Ministry of Education, Culture and Science (2018), "The Netherlands – National Plan Open Science (NPOS)", case study for OECD project on enhanced access to data, https://community.oecd.org/servlet/JiveServlet/downloadBody/149103-102-1-263392/The%20Netherlands%20-%20case%20study%20for%20open%20access%20to%20data%20for%20STI%20-%20for%20OECD.....pdf.

Norwegian Ministry of Education and Research (2018), "Case study of Norway's national strategy on access to and sharing of research data", case study for OECD project on enhanced access to data, https://community.oecd.org/servlet/JiveServlet/downloadBody/149046-102-1-263335/Case%20study%20of%20Norway%E2%80%99s%20National%20strategy%20on%20access%20to%20and%20sharing%20of%20research%20data.pdf.

Norwegian Ministry of Local Government and Modernisation (2016), "Digital agenda for Norway: Digitisation vital for welfare and jobs", press release, April, https://www.regjeringen.no/en/aktuelt/digital-agenda-for-norway-digitisation-vital-for-welfare-and-jobs/id2484184/ (accessed on 2 March 2020).

OECD (2019), *Enhanced Access to and Sharing of Data: Reconciling Risks and Benefits for Data Re-use across Societies*, OECD Publishing, Paris, https://doi.org/10.1787/276aaca8-en.

OECD (2018), *Enhanced Access to Publicly Funded Data for Science, Technology and Innovation*, webpage, OECD, Paris, https://community.oecd.org/community/cstp/enhanced-data-access (accessed on 9 January 2020).

OECD (2015), "Making open science a reality", *OECD Science, Technology and Industry Policy Papers*, No. 25, OECD Publishing, Paris, http://dx.doi.org/10.1787/5jrs2f963zs1-en.

OECD (2006), *Recommendation of the Council concerning Access to Research Data from Public Funding*, OECD, Paris, https://legalinstruments.oecd.org/en/instruments/OECD-LEGAL-0347 (accessed on 27 February 2020).

Office of National Statistics UK (n.d.), "Secure research service", webpage, https://www.ons.gov.uk/aboutus/whatwedo/statistics/requestingstatistics/approvedresearcherscheme.

Philipps, A. and S. Bodmann (2018), "Case Study Germany: The National Research Data Infrastructure / Nationale Forschungsdateninfrastruktur (NFDI)", case study for the OECD project on enhanced access to data, https://community.oecd.org/servlet/JiveServlet/downloadBody/149010-102-1-263238/20190125%20Case%20Study_Germany_NFDI.pdf.

Research Council UK (2012), "Common principles on data policy", webpage, UK Research and Innovation (UKRI), https://www.ukri.org/funding/information-for-award-holders/data-policy/common-principles-on-data-policy/ (accessed on 3 October 2019).

Shin, E. (2018), "Korean case report on enhanced access to research data", case study for OECD project on enhanced access to data, https://community.oecd.org/servlet/JiveServlet/downloadBody/141310-102-4-263210/korean%20case%20report.pdf.

Spanish Ministry of Economy, Industry and Competitiveness (2017), *Plan Estatal de Investigación Científica y Técnica y de Innovación*, www.ciencia.gob.es/stfls/MICINN/Prensa/FICHEROS/2018/PlanEstatalIDI.pdf (accessed on 27 February 2019).

Steering Committee of the National Science and Technology Council (2018), *Strategies to Promote Sharing and Use of Research Data for Innovative Growth*, Ministry of Science and ICT, https://www.pacst.go.kr/jsp/post/postCouncilView.jsp?post_id=1096&board_id=11&etc_cd1=COUN03#this (in Korean).

Swan, A. (2012), *UNESCO Policy Guidelines for the Development and Promotion of Open Access*, UNESCO, Paris, http://unesdoc.unesco.org/images/0021/002158/215863e.pdf.

The Royal Society (2012), "Science as an Open Enterprise", webpage, https://royalsociety.org/topics-policy/projects/science-public-enterprise/ (accessed on 2 March 2020).

Tramte, P. (2018), "Case study on research data management and openness in Slovenia", case study for the OECD project on enhanced access to data, https://community.oecd.org/servlet/JiveServlet/downloadBody/141328-102-1-248464/Slovenia.pdf.

Treasury Board of Canada Secretariat, Open Government Team (2018), "Case study: Canada's open government portal", case study for the OECD project on enhanced access to data, https://community.oecd.org/servlet/JiveServlet/downloadBody/149007-102-1-263226/Canada%20OECD%20Case%20Study%20-%20Open%20Government%20Portal.pdf.

Tuomi, L. (2016), "Impact of the Finnish Open Science and Research Initiative (ATT)", https://www.doria.fi/bitstream/handle/10024/127285/ATT_impactreport_final.pdf?sequence=2&isAllowed=y (accessed on 15 February 2020).

US Government (2019), "Public access to Federally funded research in the United States", case study for the OECD project on enhanced access to data, https://community.oecd.org/servlet/JiveServlet/downloadBody/149047-102-1-263336/USA%20ANNEX%20to%20ENHANCED%20ACCESS%20TO%20PUBLICLY%20FUNDED%20DATA%20FOR%20SCIENCE_Final4secretariat.pdf.

UK Research and Innovation (2016), *Concordat on Open Research Data*, https://www.ukri.org/files/legacy/documents/concordatonopenresearchdata-pdf/ (accessed on 26 February 2020).

VSNU (2018), "The road to 2020", webpage, www.vsnu.nl/Roadmap-open-access-2018-2020-English/the-road-to-2020.html (accessed on 4 October 2019).

Woolfrey, L. (2018), "The DataFirst Research Data Repository", case study for OECD project on enhanced access to data, https://community.oecd.org/servlet/JiveServlet/downloadBody/149008-102-2-264196/20190228-oecd-case-study-sa-df-v2.pdf.

Notes

[1] In the Slovenian context, scientific information comprises both scientific publications and data.

[2] Federal Ministry of Education and Research and the governments of the Federal States (*Länder*) – Joint Science Conference.

[3] Argentina: Ministry of Science, Technology and Productive Innovation; European Union: European Commission – DG Research; Finland: Ministry of Education and Culture; Korea: MSIT; Norway: Ministry of Education and Research; Slovenia: Ministry of Education and Research, Spain: Ministry of Science, Innovation and Universities.

[4] Mexico: CONACYT; Sweden: SRC.

[5] An umbrella organisation of government research institutes.

[6] open.canada.ca.

[7] The Open Maps section of the open government portal is the public interface of the Federal Geospatial Platform. NRCan led extensive efforts to consolidate, standardise, and improve access to geospatial data.

They provide data via mappable data services, much of which is science and research based. Canada is a global leader in geospatial data.

[8] www.vsnu.nl.

[9] https://knaw.nl/en.

[10] www.nwo.nl/en.

[11] www.vereniginghogescholen.nl/english.

[12] www.hetpnn.nl.

[13] www.kb.nl/en.

[14] www.surf.nl/en.

[15] www.nfu.nl/english.

[16] www.zonmw.nl.

[17] www.go-fair.org.

[18] UK Research and Innovation was formed in April 2018. It is made up of the research councils, Research England and Innovate UK; it therefore replaces HEFCE and RCUK, which were original signatories of the Concordat.

[19] https://www.opengovpartnership.org/.

[20] http://biodiversidad.co/.

[21] https://www.sibcolombia.net/acceso-abierto/.

[22] https://www.sibcolombia.net/servicios/.

[23] https://www.sibcolombia.net/el-sib-colombia/.

4 Main policy gaps hindering access to data

This chapter uses as a starting point the OECD *Recommendation of the Council concerning Access to Research Data from Public Funding* and the *OECD Principles and Guidelines for Access to Research Data from Public Funding*. A survey conducted by the OECD Committee for Scientific and Technological Policy (CSTP) in 2017 investigated the continued relevance of those principles, as well as potential additional principles for the future. In March 2018, a joint CSTP-OECD Global Science Forum workshop was held under the title: "Towards new principles for enhanced access to public data for science, technology and innovation". The workshop brought together 30 experts from government bodies, private companies, academia and non-governmental entities to take stock of current policy practices and discuss future policy needs to support enhanced access to data.

Further, the CSTP produced specific case studies of policies that illustrated good practice in policy making promoting enhanced access to data.

Relevance of the 2006 OECD *Recommendation of the Council concerning Access to Research Data from Public Funding*

The OECD *Recommendation of the Council concerning Access to Research Data from Public Funding* (OECD, 2006) (hereafter "the Recommendation") is based on a set of underlying principles (Box 4.1).

> **Box 4.1. Principles contained in the OECD *Recommendation of the Council concerning Access to Research Data from Public Funding***
>
> The principles can be summarised as follows:
>
> - **Openness**: open access to research data from public funding should be easy, timely, user-friendly and preferably Internet-based.
> - **Flexibility**: flexibility requires taking into account the rapid and often unpredictable changes in information and communication technologies (ICTs); the characteristics of different research fields; and the diversity of research systems, legal frameworks and cultures in each member country.
> - **Transparency**: information on research data and data-producing organisations, and documentation on the data and conditions attached to their use, should be internationally available in a transparent way, ideally through the Internet.
> - **Legal conformity**: data-access arrangements should respect the legal rights and legitimate interests of all stakeholders in the public enterprise. Access may be restricted for reasons of national security, privacy and confidentiality; trade secrets and intellectual property rights (IPRs); protection of rare, threatened or endangered species; and legal processes.
> - **Protection of intellectual property (IP)**: data-access arrangements should consider the applicability of copyright and other intellectual property laws that may be relevant to publicly funded research databases (as in the case of public-private partnerships).
> - **Formal responsibility**: access arrangements should promote rules and regulations regarding the responsibilities of the parties involved. They should be developed in consultation with stakeholders and consider such factors as the characteristics of the data, e.g. their potential value for research purposes. Data-management plans and long-term sustainability should also be considered.
> - **Professionalism**: institutional arrangements for the management of research data should be based on the relevant professional standards and values, embodied in the codes of conduct of the scientific communities involved.
> - **Interoperability**: access arrangements should consider the relevant international data standards.
> - **Quality**: the value and utility of data depend on the quality of the data themselves. Particular attention should be paid to compliance with explicit quality standards.
> - **Security**: attention should be paid to supporting the use of techniques and instruments that guarantee the integrity and security of research data.
> - **Efficiency**: a central goal of promoting data access and sharing is to improve the efficiency of publicly funded scientific research, to avoid expensive and unnecessary duplication of effort. This also involves performing cost-benefit analysis to define data-retention protocols, engaging data-management specialist organisations, and developing new reward structures for researchers and database producers.
> - **Accountability**: the performance of data-access arrangements should be subjected to periodic evaluation by user groups, the responsible institutions and research-funding agencies.
> - **Sustainability**: due consideration should be given to the sustainability of access to publicly funded research data as a key element of the research infrastructure.
>
> Source: OECD (2006), *Recommendation of the Council concerning Access to Research Data from Public Funding*, https://legalinstruments.oecd.org/en/instruments/159.

The 2017 Committee for Scientific and Technological Policy (CSTP) survey on policy practice related to access to data for science, technology and innovation (STI) (see Chapter 2) asked respondents to comment on the continued relevance of the Recommendation and implementation issues related to each principle, as well as propose potential new focus areas based on the evolving needs of stakeholders and policymakers. Figure 4.1 shows the survey results.

Interoperability ranked among the most relevant principles cited by survey respondents. Among the principles contained in the Recommendation (hereafter "the Principles"), interoperability included the explicit mention of the standards used, the adoption of best practices by professional organisations active in data collection and preservation, and the consideration of more general ICT standards.

While progress has been made in ensuring interoperability within disciplines, cross-disciplinary interoperability remains undeveloped. Interoperability is also a component of the Findable, Accessible, Interoperable and Reusable (FAIR) principles put forward by the European Union. Respondents proposed that the Recommendation provide guidance on ontologies and translation. The establishment of supranational open-science clouds, such as the European, Australian, African and National Institutes of Health (NIH) Commons in the United States, will generate a leap in findability for scientists in those regions. Going a step further, interoperability between those clouds needs to be established to develop global access.

Quality also ranked among most relevant principles. This principle comprises quality control through peer review, documenting the origin of sources, linking to original research materials and datasets, and data citation practices. Survey respondents felt that more needed to be done on overall quality assurance, by defining explicit and verifiable quality standards that could be captured quantitatively where possible. A potential future revision of the recommendation could provide guidelines for determining data quality, as well as a standard for labelling datasets with a confidence value.

Openness was cited as one of the most relevant principles by respondents. It is defined in the Recommendation as "access on equal terms for the international research community at the lowest possible cost, preferably at no more than the marginal cost of dissemination. Open access to data to research data from public funding should be easy, timely, user-friendly and preferably Internet-based" (OECD, 2006).

The European Union promotes an "open-by-default", efficient and cross-disciplinary research-data environment. It allows for proportionate limitations only in duly justified cases relating to personal-data protection, confidentiality, IPRs, national security or similar concerns (e.g. "as open as possible and as closed as necessary"). Australia's Open Government National Action Plan includes open access to data. However, respondents emphasise the need to limit openness in cases where legitimate reasons exist to keep data closed and warn against potential disincentive to data acquisition under an "open-by-default" policy. Further, they emphasise the need for "cultural change" among researchers to achieve more openness.

Transparency ranks very high in terms of relevance. The Principle defines it through the following aspects: i) information on data-producing organisations and their holdings, and documentation on available datasets; ii) dissemination of information on research-data policies to stakeholders; iii) agreements on standards for cataloguing data, and application thereof; and iv) information on data-management and access conditions, to be shared among data archives and data-producing institutions.

Sustainability is also considered highly relevant, and should be ensured throughout the successive evolutions in technology and standards. Hence, datasets need to be preserved across technology changes. However, user expectations should be managed, to ensure that they understand the scope and reuse potential of data – notably older data, which do not conform to the latest data standards. In this respect, regular evaluations of electronic infrastructures and services are needed, and the data-lifetime and deletion policy should be specified.

Security is another highly ranked principle. In the Principles, security encompassed both integrity (completeness and absence of errors) and security (protection against loss, destruction, modification and unauthorised access). Respondents see security as essential to fostering trust. They proposed the introduction of new guidelines on data provenance for repositories; versioning should be introduced to address data integrity. The guidelines should address not only the benefits, profits and advantages of enhanced access to data, but also its possible disadvantages and risks, to identify ways of overcoming them.

Figure 4.1. Assessment of the relevance of the principles from the OECD *Recommendation of the Council concerning Access to Research Data from Public Funding*

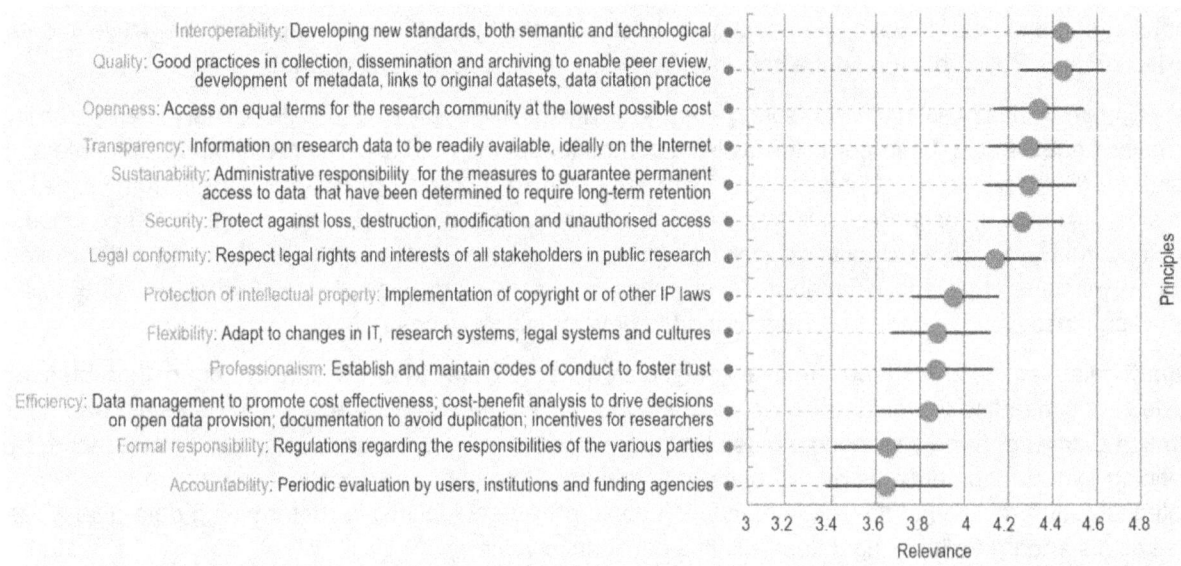

Additional principles quoted by the stakeholders but not included in the 2006 Principles:

Discoverability is crucial if data is to be re-used. Therefore there is a need to establish ontologies and appropriate semantics to enable scientist from different scientific domains to find relevant datasets.

Machine-readability should be the norm, to facilitate treatment of large quantities of data.

FAIR data principles: Findable, accessible, interoperable, reusable.

Strict rules for the financial support of open access to data: (i) responsibility for financing at international, national, regional and institutional level; (ii) definition of roles between financiers, operators and users; (iii) supervision mechanism for compliance with Open access to data rules, foresee legal sanctions.

Definition of responsibility and ownership, including legal and ethical issues.

Explicit recognition and reward system for data authorship.

Publicly funded research data treated as Commons. Licencing under Creative Commons could be used to provide a framework which ensures openness while restricting reuse as needed.

Setting an embargo period for the exclusive use of data. The embargo could vary according to the nature of the output, and provide reassurance to authors.

Implications of blockchain technologies on open access to data should be investigated, since opinions of relevance are very different in the open science field. Blockchain technology is mentioned as a potential tool which could help re-allocate private monopoly rents from innovation back into the systemic network of public collaborative science and innovation.

Notes: IP = intellectual property; IT = information technology; FAIR = Findable, Accessible, Interoperable and Reusable. The 2017 survey asked respondents to assess the relevance of the 13 principles cited in the original Recommendation (OECD, 2006) on a Likert scale (5 = very high relevance; 0 = no relevance). Responses were received from 55 organisations in 27 countries.
Source: OECD (2018a), *OECD Science, Technology and Innovation Outlook 2018*, https://doi.org/10.1787/sti_in_outlook-2018-en. The last additional principle was quoted from Soete (2016), "A sky without horizons. Reflections: 10 years after", https://www.slideshare.net/innovationoecd/soete-a-sky-without-horizons.

Legal conformity under the Recommendation includes aspects of national security, privacy and confidentiality, IPRs, protection of biodiversity and legal processes. Respondents agreed this is one of the high-priority principles to be reinforced in future revisions of the Recommendation. The current principle mentions privacy, which may need an enhanced focus going forward.

Protection of IP is considered highly relevant, and the need to include it in future revisions of the recommendation is considered moderate to high. The 2006 Recommendation considers the applicability of copyright or other IP laws that may apply to publicly funded research databases – including to data resulting from balanced public-private partnerships – while facilitating broad access to data where appropriate for public research or other public-interest purposes, duly considering the protection of commercial interests.

Some respondents stated that when partnering with private parties, the public-good nature of publicly funded research data should not be compromised – meaning that it must be freely available for the use of all – and that private parties co-operating with the public sector must acknowledge this special status and abide by these principles. They also see IP protection of public research as a complex issue and suggest that data from public research could be protected under Creative Commons, allowing data to be open, but restricting any derivative or commercial use.

Other principles are seen as having moderate to high relevance, and the need for including in any future revisions of the Recommendation is equally moderate to high (as demonstrated by their positioning in the graph on Figure 4.1):

- **Flexibility** applies to technological evolution, the evolving needs of scientific disciplines, and different countries' legal systems and cultures.
- **Professionalism** is defined in the Recommendation as the use of codes of conduct to simplify the regulatory burden on access to data, inducing mutual trust between researchers, institutions and other stakeholders involved, and setting clear rules for temporary exclusive use of data.
- **Efficiency** covers cost effectiveness, cost-benefit analysis, engagement of data-management specialists as appropriate and reward structures for researchers.
- **Formal responsibility** includes explicit rules and regulations (pertaining to authorship, producer credits, ownership, dissemination, usage restrictions, financial arrangements, ethical rules, licensing terms, liability and sustainable archiving) delineating the responsibilities of parties involved in data-related activities. If included in a future revision of the Recommendation, it should provide guidelines on the handover of responsibility for data curation from a national laboratory to the principal investigator's institution. Respondents note that some countries do not possess formal agreements on terms of access and data use, disincentivising and increasing the personal burden on researchers.
- **Accountability** implies project evaluation according to overall public-investment criteria, managing the performance of data-collection and archival agencies, monitoring the extent of reuse of datasets and the knowledge generated from reuse of existing data, and predicting future needs related to data preservation and reuse.

Respondents were also asked to quote important additional principles that are not present in the Recommendation. Below are some of their responses:

- **Responsibility and ownership**, including legal and ethical issues, should be defined.
- **An explicit recognition and reward system** for data authorship should be established.
- **Publicly funded research data** should be treated as commons: licensing under Creative Commons could be used to provide a framework that ensures openness, while restricting reuse as needed.
- **An embargo period** should be set for the exclusive use of data, to reassure authors: the embargo could vary according to the nature of the output.
- **The implications of blockchain technologies** on enhanced access to data should be investigated: blockchain is a potential tool to improve inventions' traceability, providing a way to trace the sources of innovation back into the network of public collaborative science and innovation (Soete, 2016).

Progress has been achieved in the decade since the publication of the Recommendation. Data associated with publications are now expected to be made available and the "reproducibility" agenda is getting more attention, e.g. in clinical trials. A number of research funders are now requiring open access to data.

Addressing challenges in access to research data

Nevertheless, data sharing still seems less widespread than could be expected and is limited to a small number of research fields. This seems to stem from disincentives to sharing research data, including a lack of reward or credit for sharing data; the substantial effort needed to upload and maintain data in a form that is usable by others; risks of misinterpretation or misuse; and IP and personal-data protection issues, i.e. the need to anonymise data samples. In addition, demand for shared data seems limited to a few scientific disciplines (OECD, 2015a).

The OECD Global Science Forum (GSF) identified nine challenges related to data-driven and evidence-based research in social and economic sciences (OECD, 2013). Box 4.2 synthesises those findings.

Box 4.2. Challenges to data-driven and evidence-based research identified in social sciences

A. Infrastructure and skills (or lack thereof) at the country level:

A.1 Lack of data-management planning to make datasets available for reuse

A.2 Investments in the personnel and infrastructure needed for data creation and curation.

B. Legal and regulatory barriers/challenges at the country level:

B.1 Lack of information on what data exist and lack of adoption of international standards for data documentation

B.2 Individual privacy issues and absence of a recognised framework governing the use of personal data

B.3 Legal, cultural, language and proprietary rights barriers.

C. Researchers' incentives and careers at the country level:

C.1 Incentives for researchers to ensure effective data sharing.

D. Data quality and characteristics:

D.1 Reliability, statistical validity and generalisability of different data sources

D.2 Need for greater harmonisation of social sciences data sources across countries

D.3 Increasing need for international co-ordination

D.4 Increasing need for interdisciplinary co-ordination for global challenges.

Source: Adapted from OECD (2013), "New data for understanding the human condition: International perspectives OECD Global Science Forum Report on Data and Research Infrastructure for the Social Sciences", www.oecd.org/sti/sci-tech/new-data-for-understanding-the-human-condition.pdf.

In the OECD CSTP 2017 survey, respondents were asked to assess the relevance of each of these challenges, as well as the current policy effort related to those challenges (Figure 4.2).

The most relevant of all challenges is C1 – *researcher's incentives* to ensure effective data sharing – but the consensus is that policy efforts to overcome that challenge are still weak. Cultural change is a long process. The perceived barriers and risks of enhanced access to data need to be counterbalanced by appropriate acknowledgement and reward systems. Data citation does not seem to have been widely implemented, and some respondents point out that prerequisites for it are still missing (such as data citation standards and metrics). Some countries also shared the view that open science should be embedded in evaluation systems, to ensure that researchers who provide high-quality research data (e.g. in Brazil, Canada, the European Commission, Japan and the Netherlands) are rewarded. Training in data literacy and data management is also an important aspect. Australia mentioned an interesting initiative: its Department of Employment organises an annual "GovHack" competition to reuse and remix government data, raising awareness and communicating about a "cultural shift" towards data stewardship and sharing throughout the research-data lifecycle.

Figure 4.2. Assessment of challenges related to data-driven and evidence-based research

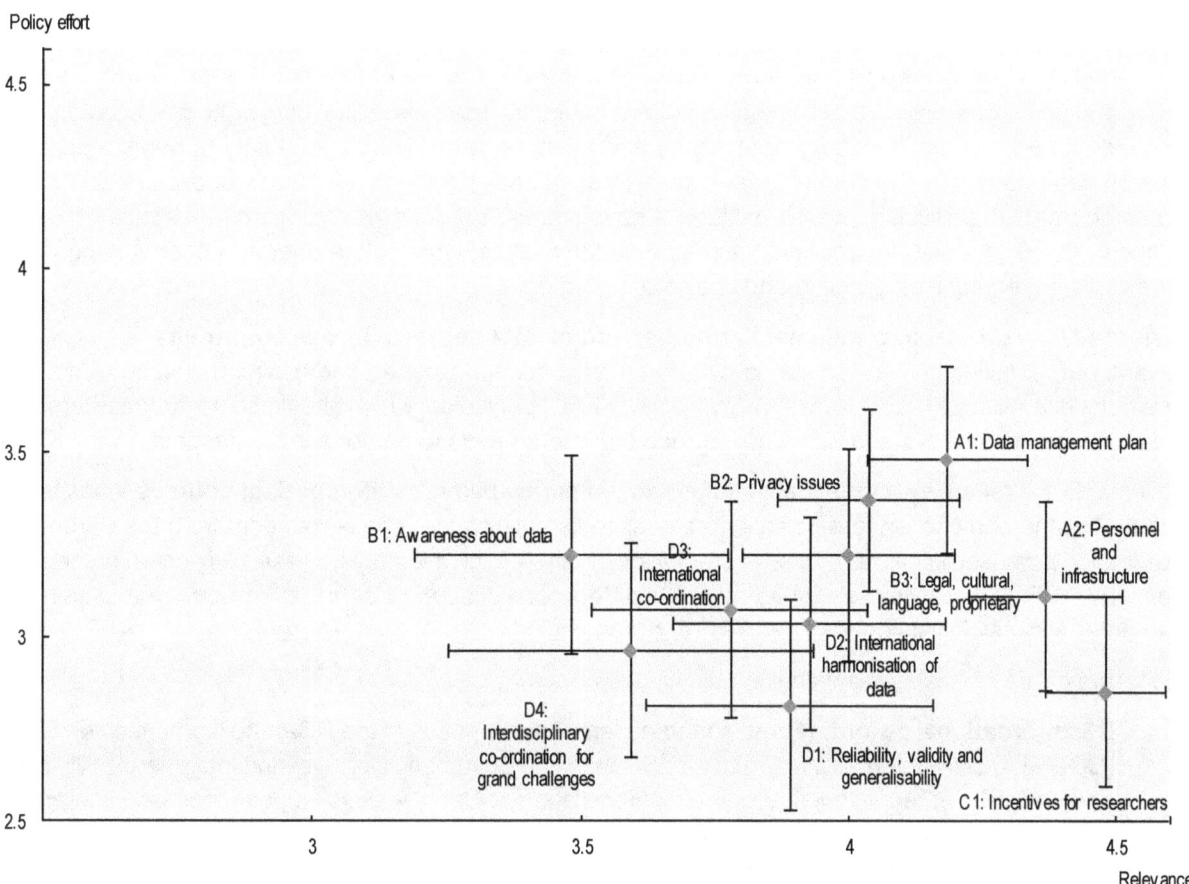

Note: An average score was computed from the responses on a Likert scale: (1 = "none"; 2 = "slight"; 3 = "moderate"; 4 = "high"; 5 = "very high"). The errors bars show the statistical error on the mean score.
Source: OECD CSTP survey results from OECD and partner delegations.

StatLink https://doi.org/10.1787/888934112614

The strongest policy effort goes into Challenge A1: the **lack of data-management planning to make datasets available for reuse**. Most governments (e.g. Australia, Canada, the Netherlands and Sweden) report they are addressing the issue by making recurrent research-funding contingent on data-sharing and data-management plans. They also quote adequate training in data-management planning as an important issue.

Another highly relevant challenge is Challenge A2: **investments in the personnel and infrastructure needed for data creation and curation**. An important initiative in this respect is the GO (Global Open) Findable, Accessible, Interoperable and Reusable (GO FAIR) initiative, led by Germany and the Netherlands, which is a proposed approach for establishing the European Open Science Cloud. The initiative rests on three pillars: i) GO CHANGE, to foster culture change, promote open science and establish reward systems; ii) GO TRAIN, to promote education and training; and (iii) GO BUILD, to build technical infrastructure (ZBW – Leibniz Information Center for Economics, 24 January 2016).

Individual privacy issues (Challenge B2) also command high policy effort and are seen as highly relevant. Governments strive to ensure a balance between maximising data sharing while ensuring the privacy and security of information, particularly through "anonymised", "non-sensitive" data. This issue is taken up in many projects and policies at the level of government as well as funding agencies and data centres. In an effort to harmonise data protection across Europe, the European Commission adopted the General Data Protection Regulation (GDPR) in 2016 (European Commission, 2016), which it enforced on 25 May 2018. There exist concerns that those stricter rules may have a negative effect an increasingly collaborative and data-intensive scientific-research sector.

Challenge B.3 – **legal, cultural, language and proprietary rights barriers** – is equally relevant, but has received slightly less policy effort, although some countries have modified copyright law accordingly. Respondents point to the necessity of clarifying and addressing the legal uncertainty of open access to research data, as well as the correct legal implementation of FAIR principles. Issues of ownership should also be addressed, particularly where institutions have created services and resources. Australia points to the necessity of harmonising legislation across data custodians, which often operate under varying legal frameworks governing the collection and use of sensitive data.

Challenge D 2 – **greater international harmonisation of data sources across countries** – is still highly relevant, but receiving only moderate policy effort. Respondents reported addressing the issue within the Research Data Alliance (RDA), as well as by applying FAIR principles, which should be the future reference for data access technical standards. They agreed that more needs to be done in this respect.

Challenge D 1 – **reliability, statistical validity and generalisability of different data sources** – has been on the receiving end of even less policy effort, despite its importance. The respondent from the European Commission proposed implementing an accreditation or certification mechanism based on agreed processes to ensure FAIR compliance, as well as establishing an accreditation or certification body to maintain an up-to-date, accessible catalogue of certified repositories.

Finally, the group of less-relevant challenges includes:

- D3: **international co-ordination**: countries report participation in the RDA, the Committee on Data (CODATA) of the International Council for Science, the Document, Discover and Interoperate Alliance, and networks of repositories (e.g. the Confederation of Open Access Repositories LA Referencia in Latin America) to work on this issue. Such co-ordination should serve to define global standards for implementing FAIR principles.
- D4: **increasing need for interdisciplinary co-ordination for global challenges**: respondents propose establishing cross-disciplinary agreements and protocols, inspired directly by relevant domain-specific needs, that will lead to specific standards. Variations across scientific disciplines, and their specific efforts to make research data open and FAIR should be respected. Developing best-practice interdisciplinary co-ordination projects could help address global challenges.
- B1: **lack of information on what data exist and lack of adoption of international standards for data documentation**: respondents recommend raising awareness of RDA standards, better implementation of data-management plans and conducting a landscape analysis of data repositories. The respondent from the European Commission proposes creating catalogues for datasets, services and standards, based on machine-readable metadata and identifiable by a common and persistent identification mechanism that will make research data findable.

Respondents were also asked to provide additional challenges not covered in Box 4.2. Some of the challenges quoted include:

- **Measurement of the status quo of data access**: general and specific indicators need to be established to measure sharing and reuse of data. Such measurement would: i) demonstrate the value added of enhanced access to data; ii) provide a basis for acknowledging and rewarding the researchers and institutions involved; and iii) help monitor the quality and sustainability of the datasets.

- **Large infrastructure solutions** to address big data nationally and internationally, with adequate governance arrangements: existing physical infrastructures need to be strengthened and new ones created to accommodate rapidly growing needs in terms of big data. Repositories featuring tools for publishing datasets are preferable to read-only portals.
- **Data reuse, data portability and interoperability**: physical infrastructures need to be complemented by internationally accepted and agreed standards, which need to be widely disseminated. An overarching recommendation on enhanced access to data could pave the way towards a more uniform political vision of these issues to trigger the needed action at the national level.
- **Funding models**: responsibility for data curation is implicitly transferred to the researcher's home institution, which may not have appropriate repositories for the specific data type. This calls for establishing mechanisms for cross-institutional and cross-border use and compensating the costs involved. Some respondents consider that publishers providing access to data at a cost is problematic.
- **Cost-benefit analysis and priority setting of enhanced access to data**: scarcity of resources implies that not all data can be made openly accessible in the short term. Hence, efforts should focus on enhanced access to the data that are most likely to provide impact, to the extent that such impact can effectively be predicted (which is not always the case). Justifying investments in infrastructure needs to be based on the value expected from enhanced access to data, which is directly related to the issue of measurement and funding models. Another related issue is selecting data for long-term preservation; this involves complex decisions about what constitutes priority data, as well as the data-preservation timespan (some research communities use data that may be centuries old).
- **Operationalise FAIR principles** in a pragmatic and technology-neutral way, encompassing equally all four FAIR dimensions: FAIR principles should be applied to all digital research objects, including data-related algorithms, tools, workflows, protocols and services. Interoperable registries of FAIR data resources should allow one to build portals to data relevant to various user needs. FAIR principles should be promoted, and the associated FAIR services should be maintained sustainably.
- **Statistical and methodological training** in use and interpretation of data, data management, and training for data standards: increasing data access and enhancing the impacts from data will require new skills. Delivering these skills will require actions from policymakers, data producers and users, and higher education institutions in the form of co-operation and partnerships, training, new education programmes and curricula – and possibly digital learning and massive open online courses. The required skills are not only technical: they include a wide range of skills in statistics, computer science, information science (e.g. for data librarians), law and other social sciences. Many countries report limited curricula to meet those skill needs.
- **Building trust** between all stakeholders, e.g. scientific communities, e-infrastructures, research infrastructures and funders, to "look outside the organisational boxes and work together": respondents suggested integrating open and FAIR access to research data in the wider context of open science (for example, the Dutch National Plan for Open Science interconnects open access, open data and reward systems).
- **Data ownership and control**: some publishers require that researchers hand over the data supporting the published article. Others offer platforms facilitating the research process, where all research elements – including annotations, methods, data and publications – can be disseminated. In the short term, this is a positive development that enhances data accessibility, but there is no guarantee that it will always be freely available.
- **Accessibility to content mining**: to the availability of research data through privately held platforms means their proprietors frequently hinder automated content mining, with the justification that these platforms present technical limitations.
- **Integrating enhanced access to data within a broader open-science approach**, including citizen science, with citizens as both providers and users of data: as a debt to taxpayers and citizens, respondents proposed transposing primary research data and scientific information into simpler representation to make it intelligible to the general public.

Synthesis of policy gaps in promoting enhanced access to data for STI

This section builds on previous analysis to synthesise the policy gaps identified as the most critical to promoting enhanced access to finance in STI, as follows:

- balancing the potential public benefits and risks of sharing – addressing privacy, confidentiality, quality and ethical issues
- technical standards and practices
- recognition and reward systems for data authors
- definition of responsibility and ownership
- business models for open-data provision
- building human capital
- exchange of sensitive data across borders.

Balancing the potential public benefits and risks of sharing: Addressing privacy, confidentiality, quality and ethical issues

The common heading of "balancing the potential public benefits and risks of sharing" – i.e. specifically addressing privacy, confidentiality, quality and ethical issues, was also a strong focus of discussions at the joint CSTP-GSF workshop "Towards New Principles for Enhanced Access to Public Data for Science, Technology and Innovation", held at the OECD headquarters in Paris on 13 March 2018. The workshop gathered 30 experts from government, the private sector, data repositories, academia, non-governmental organisations, international data networks and libraries (OECD, 2018b).

Balancing the potential public benefits and risks of sharing research data is a critical issue for data governance. Sound data governance is required to ensure trust from both data providers and users, and promote a culture of sharing, with the aim of making data "as open as possible and as closed as necessary".

Sharing data presents multiple risks, related to: i) individual privacy (e.g. in the case of clinical research data); ii) misuse (e.g. data about rare and endangered species, or rare minerals); iii) misinterpretation (particularly as concerns datasets of uncertain quality, and/or lacking the appropriate metadata); and iv) national security (e.g. data from research with potential military applications).[1] More granular data often have higher potential research value, with a concurrent increase in risk.

Providing access to personal or human-subject data is a particular challenge (OECD, 2013). Although anonymisation techniques can remove personally identifiable information from individual datasets, true anonymisation becomes very difficult as more and more data from different sources are integrated (President's Council of Advisors on Science and Technology, 2014). Moreover, the research value of personal data often stems from the ability to link them back to individual characteristics. In the United Kingdom, linking information from hospitals with the cancer-data repository and data from various screening programmes has made it possible to recommend changes in medical protocols that are likely to improve cancer survival rates. Rules and laws can be a disincentive to breaching anonymity, but the financial incentives to do so can be high in certain industries, and legal regimes are very difficult to implement across national jurisdictions.

Alongside anonymisation, informed consent is the second pillar underlying the use of personal data in research. Consent is a right recognised in many countries and enshrined in legislation, such as the recent GDPR[2] (European Commission, 2016). However, situations exist where consent to use data for specific research purposes is impossible or impractical to obtain, particularly if these purposes were not envisaged when the data were originally collected. For example, when analysing new forms of data from social networks in ways the collector had not anticipated, it might be unfeasible to go back to all the individuals concerned to ask for consent. Notably, the GDPR makes exceptions for the use of data in research, where consent is one consideration, but is not prescribed as the legal basis for data use. Recent OECD work on the subject stressed the need for properly constituted independent ethics review bodies, outlining their role in evaluating applications to access publicly funded personal data for research purposes and building trust (OECD, 2016). This recent work also emphasised the importance of public engagement in defining norms on the use of personal data in research. The approach adopted by the Australian Government, which aims to achieve value creation with open data while transparently managing risk, is one example of such an approach (Box 4.3).

Box 4.3. In my view: Trust is the key to unlocking data

The Hon. Michael Keenan MP, Minister for Human Services and Digital Transformation, Australian Government

Data are the fuel powering our new digital economy. However, news of data breaches and misuse of personal information erodes trust and leads the public to believe that data are bad or something to be feared.

If these negative perceptions become entrenched, we risk missing out on the enormous opportunities and benefits data offer to improve people's lives, help grow the economy and become more successful as a nation.

As a government, we have a responsibility to use data to make the best possible decisions to improve people's lives. In May 2018, the Australian Government announced reforms to simplify the way public data can be shared and used, and clarify accountabilities around the management of data. These reforms are made up of four components:

- a Consumer Data Right, to give Australians greater access and control over their data, to enable them to get a better deal from their bank, energy and telecommunications companies
- a National Data Commissioner, to manage the integrity and improve how the Australian Government manages and uses data
- a new National Data Advisory Council, to provide advice on ethical data use, technical best practice, and industry and international developments
- the Data Sharing and Release Act, to improve the use and reuse of data, while strengthening security and privacy protections for personal and sensitive data.

These reforms represent a tremendous opportunity to unlock national productivity. However, we will only seize this opportunity if public data are used in a safe and transparent manner, and citizens trust their privacy and security are valued and protected at all times.

To achieve that, we are working hard to secure the trust of the public at the core of our reforms.

This is the only way we can ensure the benefits of data and insights are driving effective outcomes for all people and organisations and indeed, for the entire economy and society.

Data are the fuel of growth, and trust is the key that will enable us to get ahead.

If the full benefits of enhanced access to data are to be realised, trust is required at multiple levels – not just as it relates to personal data. Power relations between individuals, institutions and countries are a critical component of trust, and need to be considered when developing data-access policies. The reality is that open research data can be more readily exploited by more advanced companies, institutions and countries that master the technology and algorithms needed to analyse extract value from the data. Less empowered stakeholders can easily be reduced to simple data providers, while the (research and monetary) value is captured elsewhere.

In order to secure public trust and accountability, the socio-economic impacts of open research data need to be monitored. Over time, such impact assessments should help society evaluate the value of open-data initiatives. The 2006 OECD Recommendation suggested considering a few core aspects for external evaluation, including overall public investments, the management performance of data collection, and the extent to which existing datasets are used and reused (OECD, 2006). This provides useful starting guidance. Nevertheless, it must be noted that such assessments are quite challenging to implement, since the methodologies are not yet well-developed and standardised. Data integration is another major opportunity. For example, New Zealand's Integrated Data Infrastructure[3] allows registered researchers to access microdata about people and households, including data on education, income and work, benefits and social services, population, health, justice and housing. Such an integrated dataset enables social-science research on issues such as the life outcomes of socially disadvantaged groups, linking their educational attainment to income, health and crime outcomes.

Balancing the potential public benefits and risks of sharing: Cross-cutting learnings from the case studies

Balancing the potential public benefits and risks of sharing is a cross-cutting theme among the 17 case studies contributed by OECD member countries and partner economies in 2018, illustrating policy practice supporting enhanced access to data for STI.

The case studies address this central issue through a consensus approach that data be "as open as possible, as closed as necessary". In those cases where the default is set to open, such as in the European Union, France and Slovenia, clear alternatives for opt-out are provided, on the condition that well-articulated reasons for opting out are formulated (European Commission, 2018; French Ministry of Higher Education, Research and Innovation, 2018; Tramte, 2018). The idea is not to put pressure on researchers to open data at all costs, but to make them think about what justifies not opening the data, and make the data accessible when no such justification exists. This is sound practice in the research-data management cycle. Although it is not yet anchored in the scientific community, it is becoming a requirement, e.g. for submitting Horizon 2020 proposals.

Some of the case studies report the development of specific agreements between data producers and users that allow the data producers to control the degree to which they share their data. At the Korea Research Institute of Chemical Technology chemical library, for example, data producers decide the degree of openness of their data (Shin, 2018). In the case of national repositories and portals, such as in Mexico, each participating institution signs an agreement that determines the degree of openness and the governance model for trust, privacy, confidentiality and ethical issues (CONACYT, 2018). In the case of Argentina's Science and Technology Information Portal, each participating institution also signs a co-operation agreement (Luchilo, D'Onofrio and Tignino, 2018).

The UK Concordat (UK Research and Innovation, 2016) addresses the issue through specific principles:

- Principle 2: there are sound reasons why the openness of research data may need to be restricted, but any restrictions must be justified and justifiable.
- Principle 5: use of others' data should always conform to legal, ethical and regulatory frameworks, including appropriate acknowledgement.

The Concordat further specifies: "Individual researchers are responsible for compliance with ethical, legal and professional frameworks, while it is the role of employers to support researchers in this through clear policies, awareness raising and providing clear advice and guidance" (Bruce, 2018).

In Canada, all heads of the government institutions are responsible for the effective, well-coordinated, and protective management of personal information in accordance with the Privacy Act and Privacy Regulations within their institutions (Treasury Board of Canada Secretariat, Open Government Team, 2018). The Canadian federal Privacy Act specifically addresses privacy protection and the right to access as follows:

- "Personal information may be collected only when it relates directly to an operating program or activity of the institution;

- Personal information must be collected directly from the person to whom it relates, with limited exceptions;
- Individuals have a right to their own personal information with limited and specific exemptions; and
- Restrictions are placed on the use and/or disclosure of personal information, subject to limited exceptions."

The Swedish case study, "Infrastructures for Register-based Research – a government commission to the Swedish Research Council" (Eriksson and Nilsson, 2018), addresses one of the most challenging issues: providing simultaneous access to two or more sensitive datasets for cross-disciplinary research. Clinical data, for example, can be combined with population statistics containing sensitive personal information, such as country of birth and citizenship. It is easy to see the value of legitimate and ethical research projects, as well as the potential risks for malevolent use. Since such research multiplies the risk of identifying individuals, researchers currently need to follow a long and complex protocol (ethical approval, harms test to evaluate sensitivity of the data, definition of scope and post-disclosure protection measures). Such a process can last many months – or even years – before the researcher can access the data. Infrastructures for Register-based Research aims to streamline this process and improve access to such sensitive data, while preserving the interests of the data providers (individual citizens). A key component is separating metadata and semantics from the sensitive data (Figure 4.3). This allows a rich dialogue between the register holder and the researcher, to define exact needs met by using non-sensitive metadata before granting access to any sensitive data.

Figure 4.3. Sweden: Separating metadata and semantics from sensitive data

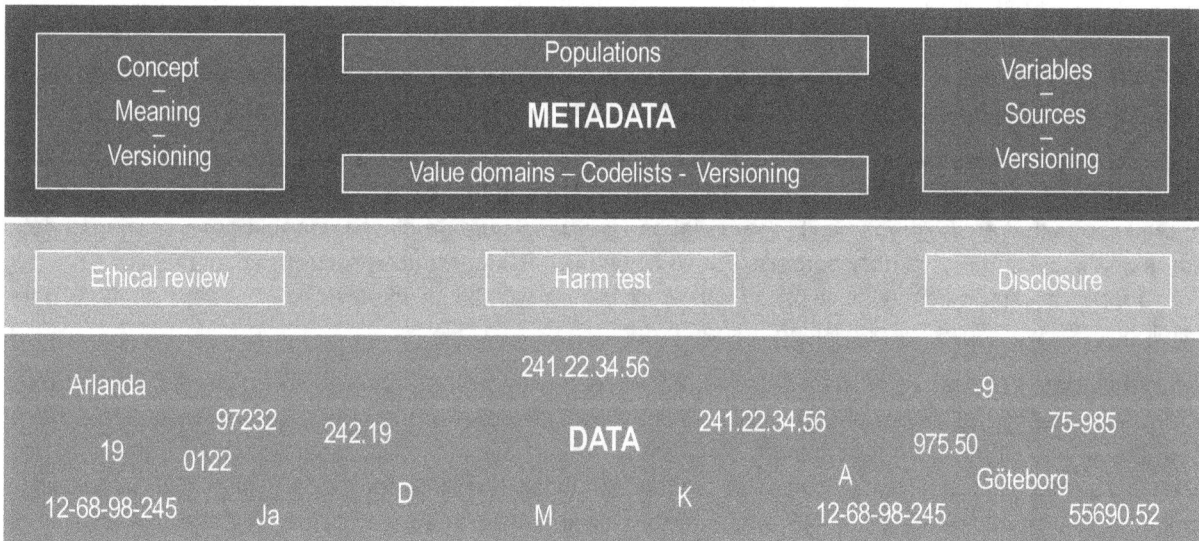

Source: Eriksson and Nilsson (2018), *OECD Case study report RUT*, https://community.oecd.org/servlet/JiveServlet/downloadBody/141329-102-1-248465/Sweden.pdf.

France has a strong ethical tradition related to the treatment of personal data, dating back to the 1978 Law on Information Technologies and Freedom, and the National Commission in Information Technologies and Freedom, whose role is to raise awareness about legal rights and obligations when dealing with personal information. This tradition is now being reinforced by the creation of a national Chief Data Officer, charged with orchestrating the circulation and reuse of public-sector data, with the goal of stimulating research and innovation while protecting the personal data and secrets safeguarded by the law (French Ministry of Higher Education, Research and Innovation, French Ministry of Higher Education, Research and Innovation, 2018).

Technical standards and practices

Technical aspects, such as dealing with discoverability/findability, machine readability and data standards, are another recurring theme in the survey findings, the workshop and the case studies.

As the volume and variety of research data increase, the resources required by data providers to make their data available, and the time invested by users to discover available data, also increase proportionally (OECD, 2015a). Insufficient information exists on what data are available, both for and from research. When data can be found, they are not always usable, because they do not conform to standards, lack metadata or are not machine-readable.

At the national scale, a large variety of institutional and domain-specific data catalogues, search engines and repositories are being established to enhance data findability (Boxes 4.1 and 4.2). At the international scale, increased efforts to co-ordinate and support global data networks are necessary (OECD, 2017) to provide the foundation for future open-science cloud initiatives that will facilitate data usage (Box 4.1).

Scientific publications are another major channel of discoverability. Many researchers first read about potentially interesting data in a journal article; the question is then how to gain access to that data. Persistent links should appear in published articles, which should also include a permanent identifier for the data, as well as the code and digital artefacts underpinning the published results. Data citation should be standard practice. Broken links or inadequate metadata are common challenges, especially as journals tend not to enforce data requirements for fear of losing good papers to competing journals. Several publishers have recently developed data journals, which can play an important role in promoting the use of published datasets. Links to data can also be included in standard publications when there are reliable sustainable services to deposit and curate data, which is already the case in several disciplines.

Formal standard-setting, through bodies such as the International Standards Organisation, is a slow iterative negotiation process that can last several years. As a result, proactive commercial or public players in a position of power can set de facto standards. One example is Google's General Transit Feed Specification, a common format for public transportation schedules and associated geographic information (OECD, 2018b).

The research community can turn this into an advantage if it takes the lead in developing appropriate standards and in so doing, consults fully with all concerned stakeholders. This is the approach taken by organisations that are helping to build the social and technical infrastructure to enable open sharing of data across national and disciplinary borders. For example, the RDA produces recommendations – which can be adopted as standards – on a broad range of issues related to interoperability, data citation, data catalogues or workflows for publishing research data (RDA, 2017).

Good metadata and the use of shared formats are essential for data interoperability and reuse. Provenance information tracks the history of a dataset and is an essential part of metadata, necessary to understand both the source of the information and the history of the dataset (it is also important for incentivising data access, as discussed in section below). In this regard, the Open Archival Information System (OAIS) reference model is of particular interest. OAIS was initially developed to archive data from space missions. It is designed to preserve information over the long term and disseminate it to a designated community, which should be able to understand the data independently, in the form in which it is preserved. OAIS covers the steps of ingesting, preserving and disseminating the data. It is universally accepted as the common language of digital preservation (Lavoie, 2014). An increasing number of repositories strive to be OAIS-compliant, which ensures the possibility of reusing data in the long term.

Discoverability/findability, machine readability and data standard: Cross-cutting learnings from the policy case studies

The Slovenian National Strategy for Open Access states: "Open research data has to be discoverable, accessible, assessable, intelligible, reusable, and, wherever possible, interoperable to specific quality standards." (Tramte, 2018) The case studies frequently cite standards as a major element that structures the initiatives and needs to be further developed. One problem is standard fragmentation, making it difficult to select the relevant standard to be implemented. The Korean case study mentions this difficulty with respect to standards for genome data (Shin, 2018). The Argentine Science and Technology Information Portal uses several standards, including the Open Archives Initiative – Protocol for Metadata Harvesting (OAI-PMH), the Dublin core and Darwin core standards; for genome data, it uses GenBank and FAST (FIX Adapted for Streaming); further it is planning to introduce DataCite (Luchilo, D'Onofrio and Tignino, 2018). The case study does not specify whether such a combination of standards gives satisfactory results. The Mexican Open Institutional Repositories Programme reports using OpenAIRE for the technical framework, OAI-PMH for harvesting processes, and DublinCore and DataCite for metadata management (CONACYT, 2018).

Sweden studied several statistical metadata standards and frameworks before finally selecting the Generic Statistical Information Model (GSIM) (HLG on Modernisation of Statistics, 2013). GSIM is a standard that specifies which types of metadata from a register should be included to sufficiently describe its detailed contents, so that metadata – rather than the sensitive data themselves – can be used to engage in dialogue with researchers, thus protecting privacy and confidentiality (Eriksson and Nilsson, 2018).

France defines data quality through completeness, up-to-datedness and reliability. These are addressed through measurement of the delay between the occurrence of an event and its publication, the availability of the infrastructure (targeted at 99.5%) and the use of open standards, facilitating reuse. France addresses findability through ScanR, a dedicated search engine for science and innovation that allows searching among datasets from 35 000 public research institutions and private enterprises in France. The search engine enables simultaneous searches of publications, projects and patent databases (French Ministry of Higher Education, Research and Innovation, n.d.).

The UK Concordat addresses the issue through specific principles:

- "Principle 7: data curation is vital to make data useful for others and for long-term preservation of data."
    - This principle quotes adherence to community-specific data formats and standards as a possible (but not exclusive) avenue to curation. Non-proprietary formats are encouraged wherever possible. Where not possible, the proprietary software needed to process the research data should be indicated. Specialised search tools and catalogues are envisioned to enhance findability.
- "Principle 8: data supporting publications should be accessible by the publication date and should be in a citeable form."
    - "[…] The dataset should be citable in itself, for example through the use of persistent identifiers, such as Digital Object Identifiers (DOIs), to ensure clarity of which exact dataset is under discussion or examination." (Bruce, 2018)

The Canadian Open Government Portal provides technical and policy guidance to individual departments and agencies, ensuring consistency, quality, accessibility and discoverability. These include: i) the Standard on Metadata;[4] ii) Open Data Release Checklist within the Open Government Guidebook; iii) Open Data Registry and User Guide; iv) Standard on Geospatial Data; and v) the Open Government Guidebook (Treasury Board of Canada Secretariat, Open Government Team, 2018).

The Netherlands, together with Germany and France, initiated the GO FAIR initiative to create an environment where data are: i) findable: easily found by humans and machines alike; ii) accessible – as open as possible, as closed as necessary; iii) interoperable – datasets need to be combinable with other datasets; and iv) reusable – it must be possible to reuse data in future research projects and process these data further (Ministry of Education, Culture and Science, 2018). As the Colombian case study points out, however, the standard is a necessary support framework, but is not a strategy for building an initiative (Escobar, Hernández and Agudelo, 2018).

Recognition and reward systems for data authors

Data sharing entails cultural change among researchers in many scientific fields. Appropriate acknowledgement and reward systems need to counterbalance the perceived barriers and risks of enhancing access to data. The emphasis on competition in research, including the way in which it is evaluated and funded, can be a strong disincentive to openness and sharing.

Researchers have incentives to publish (preferably positive) scientific results. Incentives to publish data are less developed, and usually seen as a constraint imposed by funding agencies and/or publishers. Data citation has not been widely implemented. Although the prerequisites for achieving this (e.g. standard formats and citation metrics) already exist, they are not being broadly adopted. Data activities (including related to negative results) need to be embedded in evaluation systems, to ensure that researchers who provide high-quality research data are rewarded.

Despite the progress achieved, sharing of research data remains suboptimal. In a 2016 OECD pilot survey of scientific authors, only 20% to 25% of corresponding authors had been asked to share data after publication. If asked, a significant share (30% to 50%) said they would grant access to the data, or at least undertake steps to grant it; about 30% of authors said they would seek to clarify the request. Depending on the discipline, 10% to 20% of authors would refuse to share data on legal grounds (Boselli and Galindo-Rueda, 2016). Authors of scientific papers are more reluctant to share their data openly than to access data from other research groups (Elsevier and CSTS, 2017).

There seems to be weakness of demand for data reuse as well. A recent study in the health sector revealed a low degree of awareness of open data (Martin, Helbig and Birkhead, 2015). At the same time, it also showed the limited usefulness of open anonymised data where researchers need individually identifiable data, indicating a need for tiered data release (as discussed in Chapter 1).

The Transparency and Openness Promotion (TOP) guidelines (TOP, 2014) recognise data citation as one of the levers for incentivising data sharing. They propose making data citation mandatory, as well as citing and referencing all datasets and codes used in a publication with a digital object identifier (TOP, 2014). The adoption by researchers of unique digital identifiers for researchers, such as the Open Researcher and Contributor ID, is also important in this context, as it would greatly simplify provenance mapping and data citation.

Adopting data citation as a standard practice, so that it can be used to incentivise and reward data sharing, also requires developing appropriate metrics for data citation. These could then be used alongside other assessment measures – such as bibliometrics – in recruitment and evaluation processes (OECD, 2018b). The approach adopted by the National Science Foundation (NSF) in the United States is an interesting example: over the past decade, the NSF has implemented an incremental strategy for accessing research data. Since 2013, datasets and publications are treated equally as products in an individual researcher's "biosketch". In 2016, the NSF added to the proposal section a requirement to discuss evidence of research products (including data) and their availability in prior NSF-funded research. In France, the newly published national Open Science Plan (French Ministry of Higher Education, Research and Innovation, n.d.) has adopted similar principles, pleading for a more qualitative (rather than purely quantitative) approach to evaluating researchers. The Open Science Plan is based on the San Francisco Declaration on Research Assessment, which calls for a more holistic evaluation of scientists that considers all their research outputs, including data and software (DORA, 2012).

Although recognising data citation and data products (such as datasets and databases offering enhanced or open access) in academic evaluation processes may incentivise researchers, it does not necessarily value the critical contribution of data stewards. These are the people who curate and manage data, and ensure their long-term availability and usability. Career paths for this cohort of data professionals (which include both data scientists and researchers) are unclear. Mechanisms to assess their performance should be distinct from the evaluation mechanisms applied to researchers, but should be linked to the data they manage. New measures, incentives and reward systems will be required for data stewards.

Going forward, possible policy measures to incentivise and promote data sharing by researchers include:

- developing new indicators/measures for data sharing, and incorporating them into institutional-assessment and individual researcher-evaluation processes
- promoting the use of unique digital identifiers for individual researchers and datasets, to enable citation and accreditation
- developing attractive career paths for data professionals, who are necessary to the long-term stewardship of research data and the provision of services.

Recognition and reward system for data authors: Cross-cutting learnings from the policy case studies

Most of the initiatives mention the very strong cultural barrier to sharing data among the researcher community. They cite insufficient skills for data management among researchers and insufficient resources to perform the additional workload of cleaning and curating the data for others to reuse. Above all, they highlight the perceptions of risks related to making the data available, including risks of scientific competition (another researcher acting more quickly to analyse and publish valuable results) and accrued risk to professional reputation (easier verification may increase the likelihood of uncovering errors in a researcher's analysis). To overcome these barriers, specific recognition and reward systems are needed to create incentives for data sharing.

The only initiatives that mention the inclusion of data sharing as a potential criterion for assessing scientists are the National Open Access Strategy in Slovenia, the National Plan Open Science (NPOS) in the Netherlands, and the UK Concordat:

- The Slovenian strategy states that: "research data that has undergone the scientific judgement and has been as such deposited at the authorised data centre is recognised as a scientific publication in the evaluation of the results of the programme or project". The action plan (Activity VI.1) envisages recognition of research data in research evaluation (Tramte, 2018).

- The Netherlands' NPOS (Signatories of the NPOS, 2017) states that: "[o]pen science invites a broader set of evaluation criteria than just research output and research quality, including, for example, the quality of education, valorisation, leadership and good data stewardship". Moreover, the Platform for Open Science has a working group on "Researcher recognition and rewarding". The group has issued the following recommendations: i) include (realised and expected) contributions to open science as selection criteria when hiring new researchers and support staff; ii) incorporate open science into policies on the development, support, rewarding and appreciation of scientific staff; and iii) ensure that assessment of research proposals incorporates positive rewarding of a researcher or research group's open-science track record (open-access publication, FAIR data sharing, engaging societal stakeholders); and train reviewers accordingly (Ministry of Education, Culture and Science, 2018).

- In the UK Concordat (UK Research and Innovation, 2016), Principle # 5 calls for appropriate acknowledgement when using others' data, notably: "production of open research data should be acknowledged formally as a legitimate output of the research process and should be recognised as such by employers, research funders and others in contributing to an individual's professional profile in relation to promotion, research assessment and research-funding decisions. Such formal recognition should be accompanied by the development and use of responsible metrics that allow the collection and tracking of data use and impact. In general, data citations should be accorded appropriate importance in the scholarly record relative to citations of other research objects, such as publications." (Bruce, 2018)

None of the other initiatives mention any rewards for researchers who contribute access to quality data. The Canadian Open Government Portal includes an automatic dataset format rating system (out of five "stars") for each dataset resource, based on Tim Berners-Lee five-star deployment scheme for Open Data. Users are also able to rate datasets with a five maple leaf rating system. However this user rating system is not for evaluation or reward, and specific parameters are not provided to users for what they are evaluating (Treasury Board of Canada Secretariat, Open Government Team, 2018). The Korean case study explicitly states that such incentives do not exist in Korean policy making (Shin, 2018). In Mexico, the funding agency focuses on institutional rather than individual incentives; hence, grants are conditioned on the inclusion of a target number of databases in the National Repository at the end of the project (CONACYT, 2018). Conversely, the Argentine Science and Technology Information Portal reports the existence of sanctions for researchers who do not comply with the open-access policies (Luchilo, D'Onofrio and Tignino, 2018).

The French case study mentions the development of data papers as a formal vehicle for data sharing (French Ministry of Higher Education, Research and Innovation, 2018). Such data papers generate bibliography and citations in the traditional sense, and can contribute to a researchers' evaluation.

Definition of responsibility and ownership

Issues of ownership and responsibility – including IPRs – need to be considered when enhancing access to public-research data, as they can have important implications for how – and by whom – data can be used, while ensuring full respect for the rights of data owners. Data creators may not necessarily hold the ownership of the data they collect: in the case of human-subject data, for example, the participants themselves may hold those rights.

Most saliently, any IPR associated with research data, and the licensing arrangements for the use of that data, must be clearly specified. In the absence of such specification, data acquire the statutory IPR of the jurisdiction in which they are used. This may include copyright and *sui generis*[5] database rights (e.g. in Europe), as well as local laws addressing confidentiality, privacy, trade secrets, patents and competition law which can inhibit the further use of data. Such protections arise automatically, unless expressly excluded, waived or modified (Doldirina et al., 2018). An example of how in IP and data-management policies can promote open-data practice is discussed in Box 4.4 on the example of University of Capetown.

Box 4.4. In my view: Greater clarity in IP and data-management policies can promote open-data practice

Michelle Willmers, Curation and Dissemination Manager of the Global South Research on Open Educational Resources for Development (ROER4D) project, University of Cape Town (UCT), South Africa

The ability of researchers to legally share outputs arising from their work is dictated by institutional IP policies, which are in turn largely influenced by national copyright acts. In the African context, many universities have nascent policy environments, meaning that they may not have an IP policy, or it is out of date and inadequate to cover the intricacies of online content sharing – particularly as it relates to open-data transfer and publication. There are also instances in which policy environments provide conflicting or contradictory stipulations. This situation makes for confusion on the part of academics in terms of what their actual rights are in the context of data sharing; in some cases, it may lead to flagrant disregard for policies and mandates.

Both the IP Policy and the Research Data Management Policy of UCT state that research data are owned by UCT, unless otherwise agreed in research contracts. This may lead many academics to assume they do not have the legal rights to share their data, which is not the case. UCT promotes the use of Creative Commons licensing in its IP policy, and has a concerted campaign underway to promote responsible data sharing at all levels of the academic enterprise.

Possible confusion in this regard is compounded by the fact that the institutional terms of deposit for sharing data in repositories state that: "UCT grants the Principal Investigator (PI) of a research project the right to upload UCT research data supporting a publication required by a journal publisher or a funder and all UCT project data where this is a specific funder requirement, as long as the data complies with any ethics requirements (e.g. patient confidentiality, consent, etc.)".

This caveat raises questions about the rights of academics who are not operating in research contexts led by PIs, or are functioning in a context where there is no publisher or funder requirement in this regard. The fact that the caveat only exists on a website designed to promote data sharing and is not captured in any of the formal institutional policies regulating data sharing makes the institutional open-data policy landscape confusing for academics to navigate and may serve to build reluctance and confusion, rather than promote a culture of sharing where academics are certain of their legal rights.

> Grant agreements and repository deposit terms do increasingly provide exceptions and caveats to restrictive or confusing IP policies, but these agreements are often not adequately scrutinised by academics, and the lack of cohesion between institutional policies, the dictates of funding entities and the intricacies of repository terms and conditions can ultimately amplify the distrust of – and therefore the reluctance to engage with – open-data practice.
>
> National and regional initiatives to assess and revise institutional IP policies so that they are conducive to open-data sharing and form part of a set of clear, cohesive institutional stipulations would be extremely valuable in terms of promoting open-data practice, and ensuring a functional understanding of the legal and ethical aspects of the process – the uncertainty of which often inhibits academics' practice in this regard.

Legislation and other rules for managing research data are not harmonised across organisations and countries. Data custodians often operate under various legal frameworks governing the collection and use of research data (e.g. Box 4 on South Africa). In the United States, different research-funding agencies have different IPR policies. For example, the Department of Defense has an "open-by-default" policy, while Department of Energy has a cost/benefit analysis approach to data sharing assorted with a mandatory Data Management Plan, and Defense Advanced Research Projects Agency has no identified data sharing policy (EARTO, 2016). In the European Union, copyright can be claimed on data that may not be copyrightable in other jurisdictions (such as the United States),[6] with implications for the use of text and data mining in research. According to a study by Hargreaves (Hargreaves, 2011) of the UK IP framework, "[c]opyright, once the exclusive concern of authors and their publishers, is today preventing medical researchers studying data and text in pursuit of new treatments".

Considering that one of the main drivers for enhanced access data is to improve knowledge transfer and innovation, tensions between public and private-sector actors over access to research data are a concern. Enormous potential exists for combining public-research data with private-sector data (including derived from social media). Achieving this, however, requires IPR and/or licensing arrangements guaranteeing both adequate protection of legitimate commercial interests, and the openness and transparency necessary to promote reproducibility and public confidence (OECD, 2016). The OECD *Recommendation of the Council concerning Access to Research Data from Public Funding* (OECD, 2006) states:

- Under Principle D – Legal Conformity: "Data-access arrangements should respect the legal rights and legitimate interests of all stakeholders in the public research enterprise. Access to, and use of, certain research data will necessarily be limited by various types of legal requirements".
- Under Principle E – Protection of intellectual property: "Consideration should be given to measures that promote non-commercial access and use while protecting commercial interests, such as delayed or partial release of such data or the voluntary adoption of licensing mechanisms."

Definition of responsibility and ownership: Cross-cutting learnings from the policy case studies

Several case studies touch upon the issue of responsibility and ownership.

By default, French law does not allow public administration to protect IP, and thus avoids creating a barrier (unique right of data producer) to free reuse of data. An exception is made for data produced in the context of a state-owned industrial or commercial activity in competitive markets (French Ministry of Higher Education, Research and Innovation, 2018).

In the United Kingdom, the "Research Council common principles on data" state that: "Publicly funded research data are a public good, produced in the public interest, which should be made openly available with as few restrictions as possible in a timely and responsible manner" (Bruce, 2018).

In addition, the UK case study notes that data-licensing agreements can make it complex to open up research data, creating legitimate and genuine difficulties for researchers and research organisations.

The ownership and responsibility of resources on the Canadian Open Government portal remain with the publishing department or agency. Resources published on the portal are freely reusable under the Open Government License (Treasury Board of Canada Secretariat, Open Government Team, 2018).

The Korean case study presents several use cases and approaches in different disciplines. While there exists no specific policy on "ownership" of data in the chemical library, materials data is opened only after publications and patents are released; for catalyst data, the Korean Institute of Science and Technology allows data creators to withhold information until they achieve their objectives (Shin, 2018).

In Mexico, research institutions preserve "ownership" of the data, while the national registry only aggregates registries (rather than datasets) (CONACYT, 2018). However, a survey among researchers in Mexico identified the most important barriers to data sharing as "the significant lack of knowledge about copyright and publishing policies, and the unawareness of the benefits provided by a more open research model" (Rodriguez, 2016). Argentina recommends relying on Creative Commons licences for the reuse of data (Luchilo, D'Onofrio and Tignino, 2018).

Business models for providing enhanced access to data

"Enhanced" – or even "open access" – does not necessarily imply "free of charge". However, many experts agree that public-research data should ideally be free at the point of usage, as discussed during the joint CSTP-GSF workshop "Towards New Principles for Enhanced Access to Public Data for Science, Technology and Innovation", held at the OECD headquarters in Paris on 13 March 2018 (OECD, 2018b), implying that the costs of stewardship and provision are assimilated by the data provider or repository. These costs can be substantial and require long-term financial commitment, often over several decades. Ultimately, most of the funding for open research data is likely to come from the public purse, although alternative revenue streams exist for some types of data (OECD, 2017). A key question from the science policy or funder perspective is how best to allocate this funding. The answer depends on deriving a full understanding of the business models and value propositions of specific data repositories, and the networks in which they are integrated (Figure 4.4).

Such an understanding requires considering multiple factors, including the role of the repository, and national and domain contexts; the repository's development or lifecycle phase; the characteristics of the user community; and the data product required by this community (influencing the level of investment necessary to curate and enhance the data). Business models are constrained by – and need to be aligned with – policy regulation (mandates) and incentives (including funding) (OECD, 2017).

Many different kinds of data repositories provide a large variety of services, ranging from raw data to complex online analyses. Institutional repositories, national repositories, domain-specific repositories and international repositories are all components of a complex landscape. This landscape is constantly changing as valuable new data resources arise from projects and transition into longer-term sustainable infrastructures, with longer-term funding requirements. At the level of the individual research system, potential economies of scale can be obtained by centralising or federating the management of data resources; this is common practice in some fields. However, not all data can be transferred across institutional or national boundaries, for legal, proprietary or ethical reasons. Moreover, a certain amount of redundancy in the system can also present some advantages, by making it more resilient. Federated networks can provide some benefits of scale, while respecting diversity (OECD, 2017).

Even when business models are well-developed and long-term funding is identified, there exist limits on how data repositories can operate to provide FAIR access to increasing volumes of data. Priorities need to be established and choices made, e.g. between providing immediate online access or putting data into deep storage. In the case of very big data from experimental facilities (such as the Square Kilometre Array telescope), it is impossible to provide open online access to all users; thus, tiered access systems have been developed. Prioritisation and data selection will be an increasingly significant challenge in the future. Addressing this challenge will require dialogue between repositories on one hand and data providers and users on the other, as well as more systematic cost-benefit analyses (bearing in mind that data that may have little value today can be very valuable tomorrow, and today's users may be different tomorrow).

Figure 4.4. Creating a value proposition for data repositories

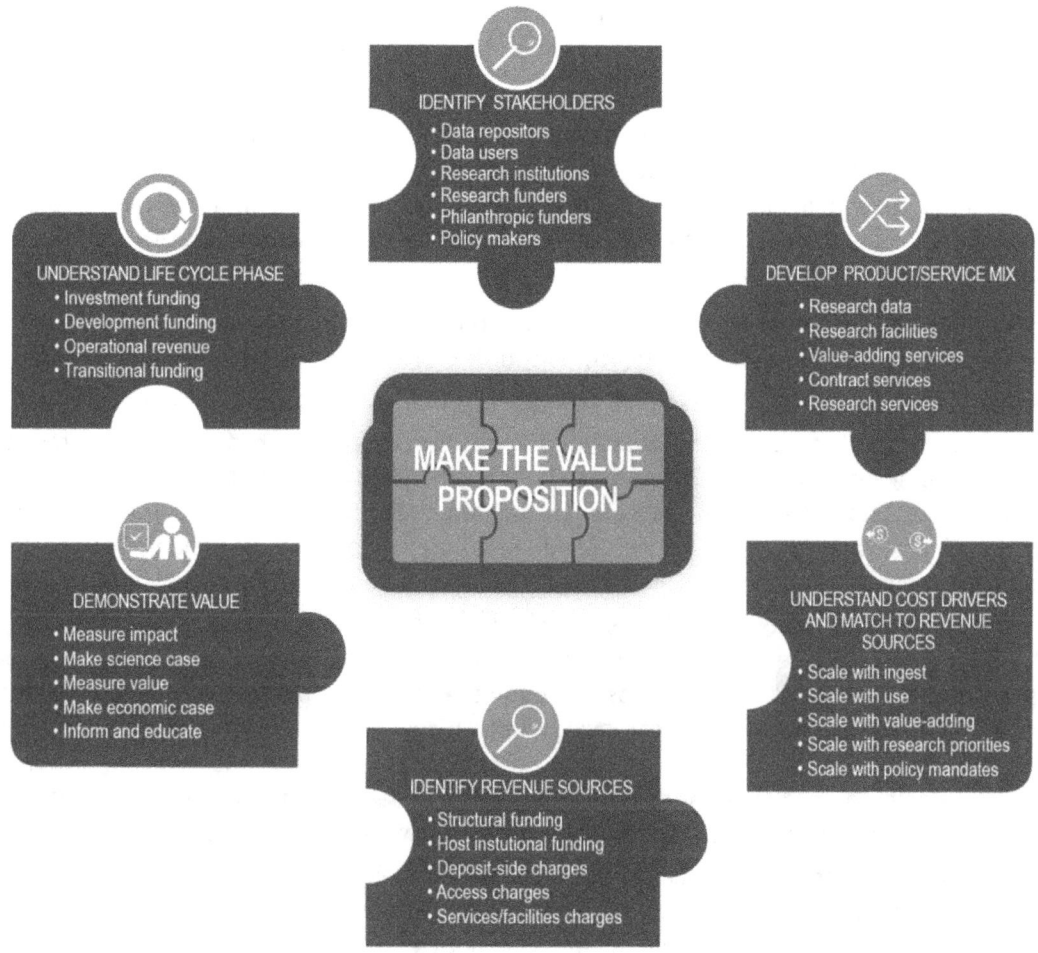

Source: OECD (2017), "Business models for sustainable research data repositories", OECD Science, Technology and Policy Papers, No. 47, https://doi.org/10.1787/302b12bb-en.

Research-data repositories and services can also be developed as public-private partnerships. Some private companies are opening their data for non-monetary gain (e.g. for recruiting, improving their image or exchanging data). For instance, medical researchers may want to combine data about people's medical history, genomics, food intake and mobility. Here, medical and genomic data may come from the public sector, whereas mobility and food data might depend on access to private-sector data. Provided IPR and ethical issues can be agreed, public-private partnerships built around such themes should be encouraged, as they can drive the development of data infrastructure and value-added services. The governance arrangements of such public-private partnerships need to be carefully designed to promote trust among all stakeholders, ensuring transparency and accountability (OECD, 2016).

Business models for providing enhanced access: Cross-cutting learnings from the policy case studies

Financing of data sharing is significant and long term, combining government, institutional and project funding. In Korea, the genomic data repository is government-funded; the chemical library is institutionally funded, complemented by project funding; and the materials pilot project runs on project funding (Shin, 2018). While the Korean pilot project on artificial intelligence data depends on project financing, public-private partnerships should be envisaged for the future, since the private sector values such data. The Colombian Biodiversity Information System, for its part, is financed by the Colombian National Ministry of Environment and Sustainable Development (Escobar, Hernández and Agudelo, 2018).

France considers data infrastructure entirely as a public investment that needs to be rendered sustainable (French Ministry of Higher Education, Research and Innovation, 2018). France is considering following the Danish example of centralised financing for data infrastructure through an interministerial platform. It also plans to establish a performance contract, with fixed targets for ministries. Finally, legal measures promoting enhanced access to data (complimentary access, reuse licences, open standards) will accelerate the data infrastructure build-up.

Principle 3 of the UK Concordat states: "Open access to research data carries a significant cost, which should be respected by all parties" (Bruce, 2018). The full text of the principle develops the business model:

- "Whilst the benefits of open research data are real and achievable, the necessary costs – for IT infrastructure and services, administrative and specialist support staff, training and for researchers' time – are significant. It is therefore vital that consideration of costs (both capital and recurrent) forms an important part of any obligation arising from the move to open research data recognising that such costs may fall outside of the defined time period of a particular project. Such costs should be proportionate to real benefits. It is recognised that the benefits and costs of open research data must be tensioned with those of the research portfolio as a whole."

- "It is UK policy that research organisations undertaking publically funded research are able to access resources for all legitimate costs through the so-called dual support system. It is therefore reasonable that appropriate costs of making research data open are met through those mechanisms whilst recognising the obligation to reduce costs through efficiency and sensible design of both obligations and infrastructure. All research-funding organisations that impose a requirement for open research data must do so in a manner that is consistent with available cost recovery mechanisms."

- "For research organisations such as Universities or Research Institutes, these costs are likely to be a prime consideration in the early stages of the move to making research data open – particularly where the required cost recovery mechanism is not yet in place. Both IT infrastructure costs and the ongoing costs of training for researchers and for specialist staff, such as data curation experts, are expected to be significant over time. Significant costs will also arise from Principle #10 regarding the undertaking of regular reviews of progress towards open access to research data. All of these costs must be balanced with the benefits to the research portfolio as a whole."

The Canadian Directive on Open Government requires each department and agency to maximise the release of government information and data of business value. Each of the federal government's Science-Based Departments and Agencies releases open data and/or information resources via open.canada.ca. Additionally, the Government of Canada has recently established the position of Chief Science Advisor of Canada. The Chief Science Advisor works to ensure that government science is fully available to the public and that government scientists can speak freely about their work. The resources on the portal are freely shared and useable under the Open Government Licence (Treasury Board of Canada Secretariat, Open Government Team, 2018).

Building human capital

Depending on the scientific domain, researchers normally have some training in data analysis, but often lack data-management skills. Users (who may be from different academic sectors, or from the private sector) do not always have the appropriate skills to interpret and analyse the data correctly. The effective operation of data repositories requires specialised skills in data curation and stewardship. Various other skills – related to ethical, legal and security concerns, as well as risk management, communication and design – should be included in any well-functioning open-data ecosystem. A lack of these skills breeds a lack of trust.

"Data science" and "data scientists" are overarching terms encompassing a wide range of skill needs. The National Institute of Standards and Technology Big Data Interoperability Framework (Volume 1)[7] defines a data scientist as "a practitioner who has sufficient knowledge in the overlapping regimes of business needs, domain knowledge, analytical skills and software and systems engineering to manage the end-to-end processes in the data life cycle." In reality, very few individuals in most scientific fields fit this definition and are leaders in each of these skill areas. Research increasingly depends on collaboration and co-operation between individuals with different data skillsets. Defining the needs and gaps for these skillsets in different scientific fields is an ongoing challenge.

Several detailed analyses exist of the data-skill requirements for science, e.g. the Data Science Framework developed by the EU-funded EDISON project[8] (Figure 4.5).

Figure 4.5. Data skills

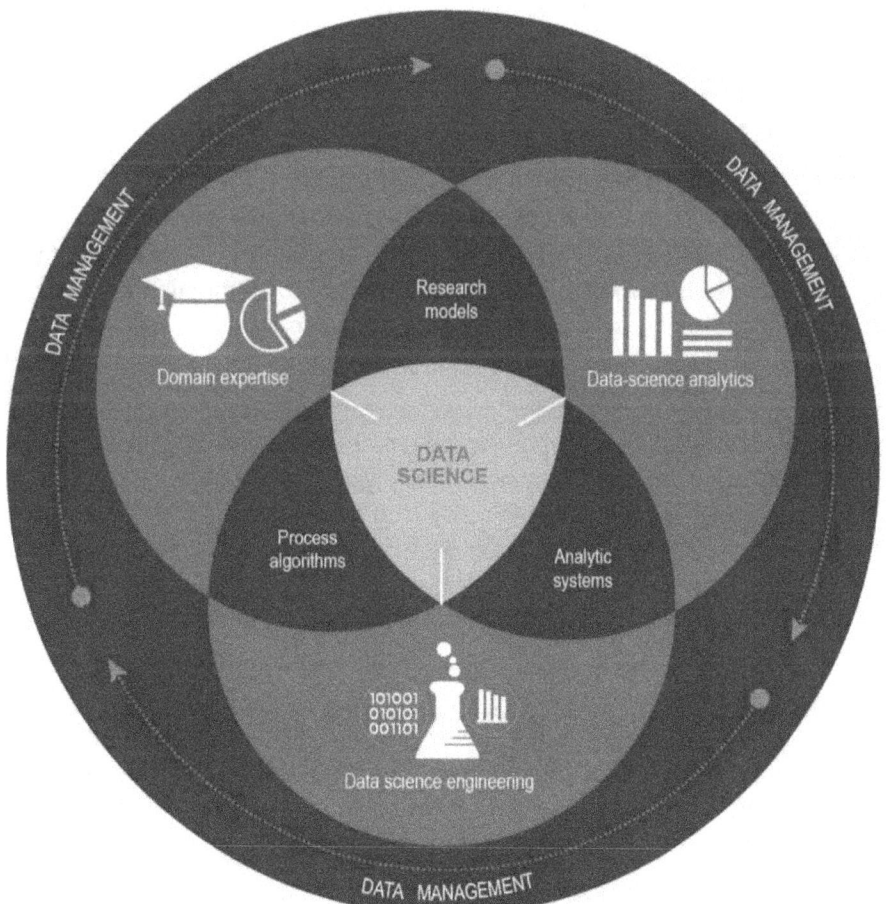

Note: This diagram illustrates the main competence groups within data science, as defined in the EDISON project: data-science analytics, data-science engineering, and domain knowledge and expertise. Data management, including curation and long-term stewardship, is sometimes classified as part of data science or as a separate competence group. These various competences need to be integrated into the different aspects of the research process, from design to experimentation, analysis and reporting.

As regards data skills, different scientific domains are equipped to varying degrees. Traditionally data-intensive fields, such as experimental physics or astronomy, are generally well-positioned (although competition for data scientists from commercial actors is affecting recruitment and retention in academia). Other areas, such as medical research, have significant skill gaps. Moreover, the additional burden of curating and stewarding data to make it available for secondary use creates a human-resource challenge that cuts across all areas of science.

Identifying skill needs and gaps across different research domains is a necessary and challenging first step. Meeting these needs is an even greater challenge; it requires retraining existing personnel (e.g. retraining librarians and archivists to perform data-stewardship functions), and provide them with the relevant new skills, as well as providing new education and training opportunities for researchers and other professional research-data support roles. Many such initiatives are already underway; they yield considerable opportunities for mutual learning across countries and scientific domains.

Data scientists are in high demand in industry, with the result that academic research competes for the best talent. An urgent need exists to develop recognition and reward structures, as well as attractive career paths, for all the specialists needed to exploit the value of data derived from public research. As in other research areas, workforce diversity will be an important determinant of success, to be considered at the outset when developing human-resource strategies for the digital research age.

Building human capital: Cross-cutting learnings from the policy case studies

Most case studies mention human capital and skills, highlighting the support researchers need to comply with open-access requirements:

- Slovenia's Open Access Strategy stipulates that public research organisations should establish support mechanisms for researchers regarding compliance with open-data requirements (Tramte, 2018).
- Belgium's DMPonline focuses exclusively on such support, having developed a specific template for data-management planning (Laureys, 2018).

One objective of the Korean strategy is to provide education and training on data skills for data scientists/engineers, and to hire data-management professionals (Shin, 2018). Argentina encourages training staff responsible for the institutional digital science-and-technology repositories (Luchilo, D'Onofrio and Tignino, 2018). Mexico mostly organises capacity-building through seminars and workshops (CONACYT, 2018).

The UK Concordat addresses human capital and institutional capabilities in Principle 9: "Support for the development of appropriate data skills is recognised as a responsibility for all stakeholders." Based on the "recognition that curating, archiving, manipulating and analysing data requires a set of skills distinct from those utilised to collect, generate, or measure the data in the first place", the Concordat calls upon research institutions to provide researcher-training opportunities in an organised and professional manner, with adequate funding from funding agencies. It further calls on institutions to ensure well-designed and sustainable career paths for data scientists in the realm of research-data management (Bruce, 2018).

Canada launched the Open Government Learning Hub[9] in 2017 to provide guidance and resources to departments. Since July 2016, the Treasury Board of Canada Secretariat Open Government team has delivered 34 events on open government with approximately 1 800 learners from at least 26 federal organisations. Canada's 2018-2020 National Action Plan on Open Government, includes a commitment to continue to promote and raise awareness and skills in the public service by continuing to build on the above (Treasury Board of Canada Secretariat, Open Government Team, 2018).

Exchange of sensitive data across borders

Data access and sharing should not be considered a "binary concept" opposing closed and open access to data. Rather, it is a continuum of openness, ranging from internal access and reuse only by the data holder (also known as closed access), through restricted (unilateral and multilateral) external access and sharing, to open access to the public (open data) as the extreme form of data sharing. Such a continuum makes it possible to address different risk-benefit trade-offs (OECD, 2015b). Low-sensitivity datasets will be candidates for open access, while more sensitive datasets can be shared on a more restricted basis with trusted and certified users.

Exchange of sensitive data across borders: Cross-cutting learnings from the policy case studies

The Swedish registry data are an example of restricted access to high-sensitivity data. Another example is the Secure Research Service provided by the UK Office of National Statistics, which provides certified researchers with access to sensitive datasets (Office of National Statistics UK, n.d.). This service provides two levels of sensitive datasets: i) very high-sensitivity "secure" datasets; and ii) intermediate-sensitivity "safeguarded data", offered under end-user licence. The secure datasets are never released to the end user; rather, they can be consulted in a "Five Safes" framework (Table 1.2). This means that only approved researchers can access the data within a specific environment, analyse it without extracting the actual sensitive data and submit the results of their research for approval. Those results will be tested to determine whether they risk disclosure. If the results are considered "safe", they will be authorised for use by the researcher; if they are considered "unsafe", the researcher will need to devise a way of further anonymising the results.

UK legislation, notably the 2017 Digital Economy Act (Her Majesty's Government, 2017), defines the different categories of personal information and the exact rules for sharing personal data, and enables the Secure Research Service to operate lawfully. However, such a service can only function on UK soil under UK legislation. It could never be provided to a researcher located in a different country, owing to a lack of international legal frameworks ensuring the same level of legal protection against misuse.

In the United States, the NIH ensures that human genomic data resulting from funded research is shared through a controlled-access mechanism, unless study participants have explicitly consented to sharing their data through unrestricted access mechanisms. Since 2007, more than 6 600 investigators from 46 countries have submitted 43 372 requests to access these data; approximately 63% of these requests were approved. On 1 November 2018, NIH updated its policy to allow unrestricted access to genomic summary results from most such studies after 1 May 2019. However, some study populations, such as those from isolated geographic regions or with rare or potentially stigmatising traits, may be made available only through restricted access (US Government, 2019).

References

Boselli, B. and F. Galindo-Rueda (2016), "Drivers and implications of scientific open access publishing: Findings from a pilot OECD international survey of scientific authors", *OECD Science, Technology and Industry Policy Papers*, No. 33, OECD Publishing, Paris, http://dx.doi.org/10.1787/5jlr2z70k0bx-en.

Bruce, R. (2018), "UK case study: The concordat on open research data", case study for the OECD project on enhanced access to data, https://community.oecd.org/servlet/JiveServlet/downloadBody/149486-102-1-264203/OECD-open-access-to-data-UK-Case-Study-v01_02%20(002).pdf.

CONACYT (2018), "Mexican open science policy – Case study: Open institutional repositories program", case study for the OECD project on enhanced access to data, https://community.oecd.org/servlet/JiveServlet/downloadBody/149104-102-1-263395/Mexico.pdf.

Doldirina, C. et al. (2018), "Legal approaches for open access to research data", *Law ArXiv Papers*, http://dx.doi.org/10.31228/OSF.IO/N7GFA.

DORA (2012), *San Francisco Declaration on Research Assessment (DORA)*, https://sfdora.org/ (accessed on 7 July 2019).

EARTO (2016), "EARTO background note: Overview of US Federal Agencies data sharing policies", https://www.earto.eu/wp-content/uploads/EARTO-Background-Note-US-Federal-Agencies-Data-Sharing-Policies-05-12-2016.pdf (accessed on 11 March 2020).

Elsevier and CSTS (2017), *Open Data: The Researcher Perspective*, https://www.elsevier.com/__data/assets/pdf_file/0004/281920/Open-data-report.pdf (accessed on 26 July 2019).

Eriksson, M. and M. Nilsson (2018), "OECD case study report RUT", case study for the OECD project on enhanced access to data, https://community.oecd.org/servlet/JiveServlet/downloadBody/141329-102-1-248465/Sweden.pdf.

Escobar, D., A. Hernández and M. Agudelo (2018), "SiB Colombia – case study: Enhanced access to public data for science, technology and innovation", case study for the OECD project on enhanced access to data, https://community.oecd.org/servlet/JiveServlet/downloadBody/149102-102-1-263390/SIB%20Colombia%20-%20Case%20Study%20open%20access-AP.pdf.

European Commission (2018), "Case study of policy initiative for open access to research data: Horizon 2020 open research data (ORD) pilot and data management plan", case study for the OECD project on enhanced access to data, https://community.oecd.org/servlet/JiveServlet/downloadBody/141323-102-1-248452/European_Commission.pdf.

European Commission (2016), *Regulation (EU) 2016/679 of the European Parliament and of the Council – of 27 April 2016 – on the Protection of Natural Persons with Regard to the Processing of Personal Data and on the Free Movement of such Data, and Repealing Directive 95/46/EC (General Data Protection Regulation)*, http://eur-lex.europa.eu/legal-content/EN/TXT/PDF/?uri=CELEX:32016R0679&from=EN (accessed on 13 September 2017).

French Ministry of Higher Education, Research and Innovation (n.d.), *Explore the World of French Research and Innovation with ScanR*, search engine, https://scanr.enseignementsup-recherche.gouv.fr/ (accessed on 27 February 2020).

French Ministry of Higher Education, Research and Innovation (2018), "Ouverture des données publiques et de recherche", case study for the OECD project on enhanced access to data, https://community.oecd.org/servlet/JiveServlet/downloadBody/141327-102-2-263394/French%20Case%20Study_v13.pdf.

Hargreaves, I. (2011), *Digital Opportunity: A Review of Intellectual Property and Growth*, https://orca.cf.ac.uk/30988/1/1_Hargreaves_Digital%20Opportunity.pdf (accessed on 4 July 2019).

Her Majesty's Government (2017), "Digital Economy Act 2017", United Kingdom, www.legislation.gov.uk/ukpga/2017/30/part/5/chapter/5/enacted (accessed on 5 October 2019).

Laureys, E. (2018), "Belgian case study on open access to data: DMP Belgium consortium", case study for the OECD project on enhanced access to data, https://community.oecd.org/servlet/JiveServlet/downloadBody/141322-102-1-248450/Belgium.pdf.

Lavoie, B. (2014), "The Open Archival Information System (OAIS) reference model: Introductory guide (2nd edition)", *Digital Preservation Coalition Watch Series*, October, http://dx.doi.org/10.7207/twr14-02 (accessed on 11 March 2020).

Luchilo, L., M. D'Onofrio and M. Tignino (2018), "Case study: The Argentine science and technology information portal", case study for the OECD project on enhanced access to data, https://community.oecd.org/servlet/JiveServlet/downloadBody/141311-102-1-248448/Argentina.pdf.

Martin, E., N. Helbig and G. Birkhead (2015), "1) Opening health data: What do researchers want? Early experiences with New York's Open Health Data Platform", *Journal of Public Health Management and Practice*, Vol. 21/5, pp. E1-E7, https://doi.org/10.1097/PHH.0000000000000127.

Netherlands Ministry of Education, Culture and Science (2018), "The Netherlands – National Plan Open Science (NPOS)", case study for OECD project on enhanced access to data, https://community.oecd.org/servlet/JiveServlet/downloadBody/149103-102-1-263392/The%20Netherlands%20-%20case%20study%20for%20open%20access%20to%20data%20for%20STI%20-%20for%20OECD.....pdf.

OECD (2018a), *OECD Science, Technology and Innovation Outlook 2018*, OECD Publishing, Paris, https://doi.org/10.1787/sti_in_outlook-2018-en.

OECD (2018b), "Enhanced Access to Publicly Funded Data for Science, Technology and Innovation", webpage, OECD, Paris, https://community.oecd.org/community/cstp/enhanced-data-access (accessed on 9 January 2020).

OECD (2018c), *OECD Expert Workshop on Enhanced Access to Data: Reconciling Risks and Benefits of Data Re-Use*, webpage, OECD, Paris, http://www.oecd.org/internet/ieconomy/expert-workshop-enhanced-access-to-data-reconciling-risks-and-benefits-of-data-re-use.htm (accessed on 11 March 2020).

OECD (2017), "Business models for sustainable research data repositories", *OECD Science, Technology and Industry Policy Papers*, No. 47, OECD Publishing, Paris, https://doi.org/10.1787/302b12bb-en (accessed on 9 March 2020).

OECD (2016), "Research ethics and new forms of data for social and economic research", *OECD Science, Technology and Industry Policy Papers*, No. 34, OECD Publishing, Paris, http://dx.doi.org/10.1787/5jln7vnpxs32-en.

OECD (2015a), "Making open science a reality", *OECD Science, Technology and Industry Policy Papers*, No. 25, OECD Publishing, Paris, http://dx.doi.org/10.1787/5jrs2f963zs1-en.

OECD (2015b), *Data-Driven Innovation: Big Data for Growth and Well-Being*, OECD Publishing, Paris, https://dx.doi.org/10.1787/9789264229358-en.

OECD (2013), "New data for understanding the human condition: International perspectives", OECD Global Science Forum Report on Data and Research Infrastructure for the Social Sciences, OECD, Paris, www.oecd.org/sti/sci-tech/new-data-for-understanding-the-human-condition.pdf (accessed on 12 September 2019).

OECD (2006), *Recommendation of the Council concerning Access to Research Data from Public Funding*, OECD, Paris, https://legalinstruments.oecd.org/en/instruments/OECD-LEGAL-0347 (accessed on 27 February 2020).

Office of National Statistics UK (n.d.), "Secure research service", webpage, https://www.ons.gov.uk/aboutus/whatwedo/statistics/requestingstatistics/approvedresearcherscheme.

President's Council of Advisors on Science and Technology (2014), "Report to the President: Big Data and Privacy: A Technological Perspective", The White House, Washington DC, United States, https://bigdatawg.nist.gov/pdf/pcast_big_data_and_privacy_-_may_2014.pdf (accessed on 27 June 2019).

RDA (2017), "All recommendations and outputs", webpage, Research Data Alliance, https://www.rd-alliance.org/recommendations-and-outputs/all-recommendations-and-outputs (accessed on 25 September 2019).

Rodriguez, C.E. (2016), *Encuesta a los investigadores en el SNI 2015. Módulo: Acceso abierto a la información científica*, report, Foro Consultivo Científico y Tecnológico, A.C., Mexico City, www.foroconsultivo.org.mx/libros_editados/acceso_abierto_a_la_investigacion_SNI_1.pdf.

Shin, E. (2018), "Korean case report on enhanced access to research data", case study for the OECD project on enhanced access to data, https://community.oecd.org/servlet/JiveServlet/downloadBody/141310-102-4-263210/korean%20case%20report.pdf.

Soete, L. (2016), "A sky without horizons. Reflections: 10 years after", keynote presentation at the OECD Blue Sky Forum, Ghent, https://www.slideshare.net/innovationoecd/soete-a-sky-without-horizons (accessed on 28 February 2020).

TOP Guidelines Committee (2014), *Transparency and Openness Promotion (TOP) Guidelines*, Open Science Framework, Centre for Open Science, https://osf.io/ud578/?_ga=2.55971502.408225042.1583516285-1827136234.1583516285 (accessed on 9 March 2020).

Tramte, P. (2018), "Case study on research data management and openness in Slovenia", case study for the OECD project on enhanced access to data, https://community.oecd.org/servlet/JiveServlet/downloadBody/141328-102-1-248464/Slovenia.pdf.

Treasury Board of Canada Secretariat, Open Government Team (2018), "Case study: Canada's Open Government Portal", case study for the OECD project on enhanced access to data, https://community.oecd.org/servlet/JiveServlet/downloadBody/149007-102-1-263226/Canada%20OECD%20Case%20Study%20-%20Open%20Government%20Portal.pdf.

UNECE High-Level Group for the Modernisation of Official Statistics (HLG-MOS) (2013), *Generic Statistical Information Model (GSIM) Specification* – webpage, https://statswiki.unece.org/display/gsim/GSIM+Specification.

UK Research and Innovation (2016), "Concordat on Open Research Data" case study for the OECD project on enhanced access to data, https://community.oecd.org/servlet/JiveServlet/downloadBody/149047-102-1-263336/USA%20ANNEX%20to%20ENHANCED%20ACCESS%20TO%20PUBLICLY%20FUNDED%20DATA%20FOR%20SCIENCE_Final4secretariat.pdf (accessed on 26 February 2020).

US Government (2019), "Public access to Federally funded research in the United States", case study for the OECD project on enhanced access to data, https://community.oecd.org/servlet/JiveServlet/downloadBody/149047-102-1-263336/USA%20ANNEX%20to%20ENHANCED%20ACCESS%20TO%20PUBLICLY%20FUNDED%20DATA%20FOR%20SCIENCE_Final4secretariat.pdf.

ZBW – Leibnitz Information Center for Economics (24 January 2016), "GO-FAIR: A member states-up strategy for the EOSC implementation", ZBW MediaTalk blog, https://www.zbw-mediatalk.eu/2017/01/go-fair-a-member-states-up-strategy-for-the-eosc-implementation/ (accessed on 28 February 2020).

Notes

[1] Risks related to intellectual property will be discussed in the section "Definition of responsibility and ownership" below.

[2] Regulation 2016/679 defines "consent" of the data subject as "any freely given, specific, informed and unambiguous indication of the data subject's wishes by which he or she, by a statement or by a clear affirmative action, signifies agreement to the processing of personal data relating to the data subject."

[3] http://archive.stats.govt.nz/browse_for_stats/snapshots-of-nz/integrated-data-infrastructure.aspx.

[4] The metadata used on open.canada.ca is aligned to the Government of Canada Standard on Metadata, but a specific Open Government Metadata Element Set was created, which is much more robust and comprehensive than the existing Standard. The Element Set is treated internally as official policy, and it enables Canada to have a robust and consolidated search for open data, as well as open information resources. Also, since Canada added schema.org microdata, those records were part of a pilot developed by google (Google Dataset Search), which consolidates open data records from various repositories to develop a true "federation" of open data.

[5] "Sui generis" means "of its own kind" or unique.

[6] The US Copyright Act applies only to the expression of the work, but does not extend to an idea, procedure, concept or discovery, while e.g. in Germany copyright protects the author in his intellectual and personal relationship to the work and in respect of the use of the work. A detailed discussion to be found in (Doldirina et al., 2018).

[7] https://bigdatawg.nist.gov/V1_output_docs.php.

[8] http://edison-project.eu/.

[9] https://open.canada.ca/en/learning-hub.

5 The future of access to data for science, technology and innovation

This chapter draws conclusions from the preceding work and proposes possible ways forward for the future of access to data for science, technology and innovation. It starts by summarising the policy issues identified and infers implications for policy makers. It concludes by providing potential scenarios and a view of potential future developments in this policy field.

Policy issues and implications

Issues identified in enhancing access to data for STI

The analytical work in the previous chapters has identified the following main policy issues:

i) **Balancing the benefits of data sharing with the risks.** "As open as possible, as closed as necessary" is gradually replacing the "open-by-default" mantra associated with the early days of the open-access movement. Opening data can provide benefits in advancing the science, technology and innovation (STI) agenda, but these need to be balanced against the issues of cost, privacy, security and malevolent uses. A staged approach should be used to enhance access to data, including sharing within communities of certified users, adapting the degree of certification of users to the sensitivity of data and creating safe environments where certified users can access sensitive datasets in controlled environments.

ii) **Technical standards and practices** – keeping up with the pace of technological progress. Applying the findability, accessibility, interoperability and reuse (FAIR) principles depends on developing and adopting common technical frameworks. The challenge is that technology development is now far outpacing standard-setting, creating regulatory gaps. Implementation of FAIR principles is an important initiative to close this policy gap.

iii) **Defining responsibility and ownership.** Intellectual property right (IPR) protection is a basic condition for incentivising innovation. However, advances in technology can provide opportunities for new methodologies, such as text and data mining (TDM). Copyright regulation which excludes temporary copies of text for the sole purpose of TDM can represent an impediment to research and innovation In the case of public-private partnerships, policy objectives should be clearly defined to expressly allow or forbid private ownership over the data derived from publicly funded research.

iv) **Incentives and rewards.** Recognition and rewards are needed to encourage researchers to share data. Current academic reward systems mostly encourage the publication of scientific results and do not sufficiently value data sharing. More remains to be done to raise awareness of open-government data among researchers and enhance the appeal of sharing access to data.

v) **Business models and funding for provision of enhanced access.** Costs are most often borne by data providers, while benefits accrue to users. Although enhanced access does not necessarily mean free data, in most cases, data should be free at the point of use. What are the financing options?

vi) **Building human capital and institutional capabilities to manage, create, curate and reuse data.** Better data skills for researchers, data stewards and users are a prerequisite for advancing data sharing.

vii) **Exchange of sensitive data across borders.** Sensitive datasets can be shared on a more restricted basis with trusted and certified users. Significant barriers currently exist to providing such services across borders, owing to a lack of international legal frameworks ensuring the same levels of legal protection against misuse.

Implications for policymakers

Building on current experience and looking forward, some policy implications can be drawn for governments in each of these domains:

i) **Balancing the benefits of data sharing with the risks**
- Public data for STI should be "as open as possible, as closed as necessary". Governance arrangements are critical when accessing sensitive data.
- Institutions should develop realistic consent frameworks and set up ethics review boards, with a mandate to arbitrage in cases where obtaining consent is impossible or impractical.

- Governments should strive to enhance trust among different stakeholders, and create consensus around data sharing and reuse. The risk of privacy breaches cannot be completely avoided, but should be managed through clear and transparent procedures. Creating safe environments that are open to certified researchers in controlled environments is a step in this direction.
- Specific initiatives can be launched to support data integration, exploring ways in which data from different sources can be combined transparently across different institutions. These initiatives should explore important issues related to sensitive data, such as anonymisation and informed consent.
- Socio-economic assessments should be undertaken to monitor the impact of open research data, with specific attention to where – and to whom – benefits accrue. Steps should be taken to ensure broad access to research data, to avoid a new "digital divide".
- Solutions should be explored for data integration across sectors and disciplines (e.g. providing cross-disciplinary access to register data in Sweden).
- Governments should implement proactive campaigns to foster systemic trust in data initiatives. Measures include informing and actively engaging with stakeholders, performing risk management and mitigation, and enforcing mandatory breach notification.

ii) **Technical standards and practices – keeping up with the pace of technological progress**
- Developing and adopting community-agreed standards is critical to FAIR data. Individuals and bodies (such as the Research Data Alliance) that work in this area should be supported accordingly.
- Good metadata are critical to data interoperability and reuse. Data controllers should be encouraged to comply with standardised reference models (e.g. the Open Archival Information System).

iii) **Defining responsibility and ownership**
- Information about ownership and licensing should be contained within the metadata, and specified for all prospective data products in research-data management plans. Open-use licences should be used wherever appropriate (OECD, 2015).
- OECD member countries, including their intellectual property (IP) protection expert agencies, should carefully consider the implications of any amendments to copyright legislation and IPR regimes as they relate to access to publicly funded data for research. They should implement well-established international IP norms and promote (rather than inhibit) research and innovation in new areas, such as text and data mining, and deep learning.
- Specific arrangements could promote data sharing in public-private partnerships, while respecting the legal rights and legitimate interests of all stakeholders in the public-research enterprise.

iv) **Incentives and rewards**

Specific policies should be promoted to incentivise data sharing among researchers. Such measures could include:

- developing new indicators and metrics for data sharing, and incorporating them into institutional-assessment and individual researcher-evaluation processes
- promoting the use of unique persistent digital identifiers for individual researchers and datasets, to enable citation and accreditation
- developing attractive career paths for data professionals within the research-data ecosystem – they are necessary to the long-term stewardship of research data and service provision, but are very often attracted by private-sector careers over public-sector ones.

v) **Business models and funding**

Public data for STI is a public good. Sound policies are needed to optimise data sharing while ensuring rational use of public monies. Governments could consider:

- developing strategies and roadmaps, including long-term funding plans and business models, to build a sustainable research-data infrastructure (i.e. data repositories and services)
- exploring how public investment in research data and infrastructure can be used to leverage private investment (as well as skills and data resources), while ensuring openness and accountability.

vi) **Building human capital**

Human capital is needed to manage, create, curate and reuse data. Depending on the discipline, scientists may or may not have adequate data-management skills. Users of the data do not always have the appropriate skills for interpretation and analysis. The data stewards sometimes lack skills to apply the relevant standards. Possible measures include:

- developing a national data-skill strategy for STI, identifying specific skill gaps, as well as the education and training requirements needed to fill them
- facilitating co-operation across different education and research actors, to ensure coherence and complementarity in capacity-building activities for data skills.

vii) **Exchange of sensitive data across borders**

- International legal frameworks should be developed to ensure the same level of legal protection against misuse. Such frameworks should adopt common approaches and rules for sharing data (especially personal and other confidential data) in safe environments, facilitating exchanges across borders.

Potential future developments in this policy field

The significance of data for STI will undoubtedly continue to increase over the next decade. The volume of data produced globally amounted to 16 zettabytes (ZB) in 2016 and is expected to grow to 163 ZB by 2025 (Reinsel, Gantz and Rydning, 2017). The importance of artificial intelligence in assisting scientific discovery is also expected to grow significantly. Access to well-managed data is a key enabler of this development (Kitano, 2016).

Enhanced access to research data holds considerable promise for increasing research productivity and innovation, and developing solutions to complex societal challenges. However, realising this potential – and minimising the potential risks – will require strategic planning and policy interventions. The *OECD Recommendation of the Council concerning Access to Research Data from Public Funding* (OECD, 2006) and the more recent FAIR principles for data access provide a broad framework for policy development and co-operation across communities. Many countries have already taken up the challenge, and adopted open-science policies and/or strategies on open access to research data. At the European level, the European Commission has taken the lead in ensuring policy coherence across countries.

Beyond data as such, access sharing will need to be increasingly applied to a broader category of scientific information, including software and publications. Clearly, data are linked to publications for reasons of reproducibility, and a growing number of publishers and funders require the publication of supporting data. Over time, this trend may be extended to raw data, beyond the subset needed for immediate reproduction of results (for example, the European Commission's Data Management Pilot encourages sharing data beyond the bare minimum on a voluntary basis). Conversely, access to datasets should be linked to shared access to publications featuring the results obtained from these data.

Access to the appropriate software and algorithms is gaining importance in ensuring the correct usage and interpretation of data. This interdependence means that large volumes of data can only be analysed with appropriate algorithms, and vice versa: algorithms can only be trained through exposure to vast quantities of data.

Blockchain technology is a potential tool that could improve the traceability of inventions, providing a way of tracing the source of innovation back into the network of public collaborative science and innovation.

Successful implementation of open-data policies and strategies crucially requires establishing governance systems and processes that ensure transparency and foster trust across the research community and society at large. Mandates and incentives will need to be used judiciously to support and facilitate changes in research behaviour, without stifling creativity and innovation. Long-term investment in technical infrastructure and human capital will be required. Technical standards need to be developed, and legal and ethical concerns addressed.

Several ways forward are possible, of which Box 5.1 presents two extreme cases. It is up to policymakers to decide which scenario would best suit the national interests and to activate the levers to promote the preferred scenario.

Much needs to be done, but much is already being done. Understandably, policy intervention focuses on exploiting the exciting opportunities created by enhanced access to research data. Enhanced access to data can help address issues related to the reproducibility and accountability of scientific research, provide solutions to pressing socio-economic challenges and unite the global scientific community around these issues. Looking to the future, however, it is also important to consider and mitigate the potential risks.

The advent of data-driven science coincides with a crisis of confidence in science and the advent of the "post-truth" era. Opening up public-research data means that new actors will be able to analyse and interpret the data from their own perspectives, and not necessarily with the critical objectivity expected from scientists. The old adage "if you have enough data, you can prove anything" is not unfounded.

In the new world of open science, the scientific community will need to work rigorously, clearly communicating the scientific method and limitations of its analyses, and engaging in honest discourse and dialogue with the public and policymakers. In a hyper-competitive research enterprise characterised by enormous pressure to succeed and growing hype around scientific breakthroughs, it is vital to ensure that open science and data can be trusted. Technological developments (such as blockchain) can assist in this regard. Ultimately, however, trust is a social construct, which needs to be carefully nurtured over time.

Box 5.1. Possible future ways forward for enhanced access

In a possible "best-case" scenario, trust would be earned across society, thanks to strong governance initiatives ensuring strong risk management and mitigation, elaborated in transparent consultation with stakeholders. Ethics review boards would credibly represent individual interests and arbitrage consent issues. On the technical side, strong global standards would emerge, akin to Transmission Control Protocol/Internet Protocol for Internet communication, complemented by more specialised standards for specific applications. IPR and licensing provisions would promote responsible data access and reuse and comprise a standard part of machine-readable metadata. Data citation would be ubiquitous and could become an integral part of researcher evaluation. Financing of repositories would be based on long-term infrastructure strategies and sustainable models. Finally, digital skills would be addressed through a strategic approach encompassing initial education and lifelong learning for data producers, stewards and users.

A "worst-case" scenario is also possible, in which repeated security and privacy breaches would be inadequately managed, fostering a general level of mistrust. Standards would continuously lag behind technology development, while IPRs would be insufficiently defined to support widespread data reuse. Incentives for researchers to publish their data would remain weak, and initiatives to develop data skills would be poorly designed.

www.ingramcontent.com/pod-product-compliance
Lightning Source LLC
LaVergne TN
LVHW061939070526
838199LV00060B/3876